A MIRROR TO NATURE

PETER
KNUDTSON

A MIRROR TO NATURE

Reflections on Science, Scientists, and Society

Stoddart

First published in 1991 by
Stoddart Publishing Co. Limited
34 Lesmill Road
Toronto, Canada
M3B 2T6

Canadian Cataloguing in Publication Data

Knudtson, Peter
A mirror to nature

ISBN 0-7737-2467-2

1. Life sciences. 2. Scientists - Biography. 3. Indians of North America - Social life and customs. 4. Racism. I. Title.

Q171.K59 1991 500 C91-093414-2

Some of these articles first appeared in a slightly different form in the following publications: "Harbor Seal Mother and Pup: A Primal Bond" in *Natural History*, May 1974; "The Case of the Missing Monk Seal" in *Natural History*, October 1977; "Imps of Darkness: Marine Iguanas of the Galápagos Islands" in *Oceans*, September 1980; "The Calvin Cycle: Chronicles of Maverick Berkeley Nobel Laureate Melvin Calvin" in *California Living*, March 6, 1983, reprinted with the permission of *The San Francisco Examiner* © 1989 *The San Francisco Examiner*; "Painter of Neurons: The Artistry of Spanish Nobel Prize-winning Neuroanatomist Santiago Ramón y Cajal" in *Science 85*, May 1985; "Lights, Camera, Science: The Nature of David Suzuki" in *Science 86*, August 1986; "The Heart's Lonely Hunter: Canadian Cardiac Scientist Adolfo de Bold" in *Equinox*, Copyright © 1987 Telemedia Publishing Inc.; "Ways of the Wintu" in *The Wintun Indians of California*, Naturegraph Publishers, Healdsburg, California, 1977; "Weavers of Wood: The Ancient Art of Haida Basketry" in *Natural History*, May 1982; "Rushton on Race" Copyright © 1989 Telemedia Publishing Inc.; "Redesigning Life" in *Equinox*, January/February 1988, Copyright © 1988 Telemedia Publishing Inc.; "Blaming Crime on Chromosomes: The Mystery of the Man with Too Many Y's" in *Genethics* by David Suzuki and Peter Knudtson, Stoddart Publishing Co. Limited, Toronto, 1990.

Cover design: Brant Cowie/ArtPlus Limited
Typesetting: Tony Gordon Ltd.
Illustrations from *Genethics* by Gary Cody

Printed on recycled paper in the United States of America

To my nieces Kelli and Kelly,
and nephews Randy and Clark;
and also to Byron, Alex, Juana, and Tijash,
young and able bearers
for the next generation
of precious Mayan memories

Contents

Acknowledgments

MY DEBTS, over nearly two decades as a writer, are too numerous to mention here. But I would like to express my gratitude to Bernard Taper, whose writing workshops at Berkeley have been a continuing inspiration to me, and to David Suzuki, who has placed enormous trust in my work in recent years. Others who have contributed directly or indirectly to specific items in this collection include (in alphabetical order): Sylvia Barrett, Barbara Bjorndahl, Tirthankar Bose, Annette Desmarais, Nancy Flight, Jan Kraepelien, and Andres Lopez. It was Rosemary Allenbach who first stirred in my mind the possibility of so presumptuous an undertaking. Finally, I am deeply indebted to the patient efforts of Alison Reid, my editor at Stoddart, and to Doris Cowan, who assisted her.

Introduction: On Nature Writing

Be not too tame neither, but let your own discretion be your tutor: suit the action to the word, the word to the action to hold, as 'twere, the mirror up to nature; to show virtue her own feature, scorn her own image, and the very age and body of the time his form and pressure.

Hamlet, *Act 3, Scene 2*

IT HAS ONLY BEEN quite recently — after writing magazine articles, as well as a few books, about science, medicine, and anthropology for about 15 years — that I have come to understand that in a way I'm really not a *science* writer at all. Science writers, for the most part, confine their chroniclings strictly to the so-called "objective" world of scientific discovery and knowledge. And, especially in much of my earliest efforts, I have generally played by this unwritten rule. At least I have tried to. When, on occasion, my "objectivity" as a science journalist inadvertently lapsed (permitting, for instance, a personal aside to percolate to the surface of my writing) I could always count on some distant science editor to save me from my "irrational" ways. With the patience of a well-meaning parent, he or she would prune away, usually without any comment, my every shy metaphor, emotion, or musing — in the hope that the predictably clever, dispassionate, and disembodied voice of the loyal science watcher would come through.

Only in the past decade or so (I am now 43) did it fully dawn on me that in my writing, as in other crucial areas of my personal, family, and social life, I simply *had* to begin to grant greater expression to awe, anger, joy, fear, and other such "irrational" manifestations of ordinary human experience that lie beyond and complement the mind's cool intellect and reason. This personal transformation unfortunately did not suddenly turn me into some passionate poet of science. Nor, for that matter, did it erase my every childhood trauma and hidden terror, or bring instant intimacy to my every bond. But it has, I think, helped me as a writer, and certainly as an individual, to grow a little more confident, empathetic, and complete.

If nothing else, it has made me increasingly insistent that my work be permitted to reflect my sense not just of science but also of the wider contextual world out there made up of living individuals and communities — among them some of the least powerful members of society and the biosphere; and insistent also that — whether through the choice of a word, a story's slant, or, most important, the very subject I choose to write about — my writing be allowed to express more fully (though, in some scientific circles, this seems to border on heresy) the full range of legitimate human responses to the relentless progression of the Western worldview. Insistent, finally, that my writing be allowed to serve as what I have lately come to think of as a mirror to nature — to borrow Shakespeare's apt phrase, when Hamlet offered his sage advice to a troupe of actors on how best to honestly capture the elusive realities of human lives upon the stage. Not *the* mirror, mind you, just *a* mirror — one of many — mine — however flawed or cracked or fogged by my own peculiar character and history. For, after all, *every* mirror distorts reality, if only by neatly reversing that which it reflects.

So in recent years, by my own account at least, I have slowly begun to metamorphose professionally from a *science* writer into a *nature* writer. That is, I have become a self-appointed member of what an elderly friend and former neighbor (a highly gifted American naturalist and nature writer by the name of Victor Scheffer, now in his eighties and still passionately writing) warmly assured me not long ago remains an ancient and honorable lineage of scribes. A transgenerational community of writers obsessed not simply with the blinkered visions of natural philosophy, for example, or science but, to use *Webster's* succinct definition of the word *nature*, with "the external world in its *entirety*" (emphasis mine). To my mind, that phrase grants an aspiring nature writer enormous license to explore, as here, a range of topics including darker realms (genetic folly, species extinction, or academic racism, for instance) that often tend to be considered unseemly to mainstream writers, readers, and makers of modern science.

I would hasten to add that I don't intend by so grandly titling this modest collection of my writings (previously published magazine articles, for the most part, as well as a few excerpts from books or books in progress), to harbor any lofty pretensions about its literary worth. I am a scientist by training, with a belated master's degree in biology, earned by studying the social behavior of wild harbor seals from a canvas kayak and a rickety wooden blind on intertidal mudflats of the northern California coast, after earlier abandoning a doctoral program in neurobiology in order to undertake my first tentative travels to "exotic" places. Despite a kaleidoscope of jobs — from wildlife biologist, solar-energy activist, and reluctant computer-manual writer to farm laborer (moonstruck by a young French-Canadian farmer and feminist whom I later married) in Saskatchewan wheat fields and a teacher in an alternative high school designed for young, troubled Alaskan natives — I have seldom strayed very far from, or ceased writing about, the familiar sanctuary of science and scientists. And with rare and notable exceptions, I remain convinced, science tends to breed relatively timid, unimaginative writers. (Though scientific training can without a doubt provide passage to worlds — from the microcosm of molecular genetics to the macrocosm of an expanding universe — most other writers fear to enter).

I would like to add one final note of heartfelt humility. Those of us who try to make a living by writing about science and the natural world are indebted more than most writers to the earlier unsung efforts of others — in particular to the long, arduous years of work and thought by scientists and scholars, most of whom find themselves ill equipped to communicate with a general audience. At moments when both my mood and financial status (always precarious as a freelancer) chanced to peak at the same instant, I have, on occasion, gloatingly compared this primal bond between science (or nature) writer and scientist to that between top predator and prey. When these simultaneously plummeted I was more prone, in my despair, to see it as utterly parasitic.

There is, of course, a modicum of truth in both of these rather extravagant views of the nature writer's "cruel" fate. But recently I stumbled across another much more fitting and suitably earthy metaphor. The life of a nature writer is, I have come to believe, more profitably compared to that of an ordinary earthworm — "Nature's ploughshare," as Charles Darwin, a distinguished nature writer in his own right, once so affectionately described that species. Like earthworms we are a distinctly unglamorous lot and in the short term, perhaps, too easily scorned or taken for granted. Like these slow-moving, spineless annelids, most of us spend our days in a world of semidarkness, burrowing laboriously, often blindly (in libraries and musty archives), through layer upon layer of leafy litter and organic debris — the rich detritus of lives lived far removed from our own.

In this, like worms, we are in essence little more than gluttonous digestive tracts in motion. We press forward by devouring virtually everything in our path, patiently digesting what we can. And, to make the metaphor biologically complete, there are times when the fruits of our long labors turn out to represent little more than the written equivalent of that creature's endless spoor of soft, finely textured, dirt-hued *castings* — earthworm excrement, to be blunt.

Of course, if there were no more meaning to the lives of nature's worms and writers than eternal excavation and excretion, there would seem to be little reason for either creature to exist. But as Darwin once noted in an ingenious bit of field research that still stands as a genuine contribution to nineteenth-century soil ecology, earthworms contribute significantly, however unknowingly, to the larger systems and processes of the natural world of which they, like *all* creatures, are an integral part.

In December 1842 Darwin spread a quantity of crushed chalk over a pasture not far from his home. Twenty-five years later he dug a trench across the same field to expose a neat, horizontal layer of white powder fully 20 centimeters below the ground's surface. However insignificant the aimless wanderings of a solitary individual earthworm, Darwin reasoned, the accumulated activities of a teeming population of earthworms in his pasture had over a quarter of a century gradually carpeted the initial layer of chalk particles with generations of earthworm castings. The result was a 20-centimeter-thick layer of rich, aerated, organic humus — the very stuff of fertile topsoil.

I trust my point, however strained the metaphor, is clear. Could those who scribble endlessly about the underlying interconnectedness of *all* of nature ask for more than that, like Darwin's lowly earthworms, their lifetimes of inglorious labor might, with time and in combination with the work of others, lay the groundwork for a richer, more various world?

The accumulated writings that I have selected for this book I think reflect to varying degrees these admittedly romantic views on the true nature of nature writing. And together they form a sort of mosaic — a rough, chronological record of my often erratic and still unfinished passage from science to nature writer. They stand as reflections of the times, and state of mind, in which I penned them; I have not attempted to update them.

In Part I, "On Nature," three magazines pieces serve as examples of some of my early explorations as a writer into areas of the natural world itself (namely, the natural history of marine life). As a young graduate student and field zoologist in California and Alaska, I keenly felt a visceral sense of kinship with other species that I have never lost. My biological studies fueled it, but it was kindled, and still burns, because of the gift of prolonged, direct personal experience in nature.

In Part II, "On Nature Watchers," are four profiles, written during the 1980s, on eminent scientists — an American Nobel-laureate biochemist, a well-known Japanese-Canadian geneticist, a Spanish Nobel-laureate neuroanatomist, and a pioneering Argentine-Canadian heart physiologist. These reflect my continuing fascination with the role of the individual human being in the accelerating dialogue between Western science and nature. Each of my subjects, by no accident, turns out to have been not only a gifted researcher in his field but also a genuine maverick — an original — within the larger community of scientists; and each has paid a price for his professional eccentricities and independence of thought.

In Part III, "On Native Cultures and Nature," I have included four items: two articles, an excerpt from a book on California

Indians I began at the age of 25 and a short unpublished essay. These suggest the sense I have developed over the years of the profound contribution primal, or indigenous, cultures around the world can make to the heightening of human empathy and responsibility for nature in all its manifestations. At this writing, I am in the midst of co-authoring a book that will explore in rich detail that very theme.

The book chapter and lengthy magazine article making up Part IV, "On the Nature of Human Differences," contain, I suspect, what to some readers will be the most disturbing material in this collection. It is also, in my view, perhaps the most important. It mirrors one of the darkest dimensions of human relationships with fellow human beings and the natural world — one that the West has alternately nurtured and ignored. It is our society's obsessive notion of value-laden human differences arising "biologically" from race.

For a book in progress on Guatemalan history, I have written about racist *practice* in my account of the unimaginably brutal sixteenth-century subjugation of the Maya Indians of Mesoamerica by "civilized" Europeans who, often with the full sanction of both Spain and the Catholic Church, tended to see their military and technological might (despite the extraordinary intellectual and material sophistication of the literate Maya world) as a reflection of the "natural" hierarchical order of the cosmos. Mixed-blooded descendants of the first waves of these invading *conquistadores*, still harboring some persistent modern form of this poisonous myth of racial superiority, continue to this day to hold Guatemala's vast Maya *majority* as virtual political hostages in their own traditional lands. It is Indian *apartheid* — acted out in shades of brown rather than South African black and white — pure and simple.

Racist *theory* is intellectual speculation purporting to divine a convenient hierarchy of race-based differences between human groups. Such theory has tragically too often been buttressed historically by elements of the scientific community itself. I wrote about it in the case of a highly controversial Canadian psychologist by the name of Philippe Rushton. Rushton, as you may recall, won almost instant celebrity status early in 1989 by attempting to raise once again the disturbingly familiar pseudo-scientific banner, with a few clever new wrinkles, of the supposed genetic "inferiority" or "superiority" of particular human races.

Finally, Part IV, "On Lessons from Nature," consists of one chapter from *Genethics: The Ethics of Engineering Life*, a book on moral issues arising from recent breakthroughs in modern genetics that I co-authored in 1988 with the Canadian geneticist and broadcaster David Suzuki. I met Suzuki for the first time a few years earlier, while I was preparing the profile that appears in Part II. I have also included a magazine article, published by *Equinox* in 1988, that provides a broader if by now subtly dated perspective of the same subject. These two final pieces, along with the Maya and Rushton items, are representative of some of my most recent — and, I feel, to date most emotionally complete — efforts as a nature writer, in the sense that I have previously attempted to describe. In them, as perhaps less successfully elsewhere, I have consciously struggled to hold my own hard-won and far from fully polished mirror up to nature — or at least, with a little luck, used it to cast a beam of light on a few terribly important if uncomfortable topics.

I still have much to learn.

Peter Michael Knudtson
Vancouver, B.C.
November 1990

Part I

ON NATURE

Introduction

It is seldom possible, it seems to me, to develop a deep sense of reverence for the natural world — not just for living plants and animals but also for rocks and rivers and rain storms and all the rest — without spending a good deal of time in its midst. The intimate bonds with nature that arise out of long, silent hours (including time alone) in healthy natural ecosystems are not the same as those that arise from reading a book, ambling through a city zoo, or gazing, mouth agape, at two-dimensional herds of wildebeest thundering across a TV screen, or even, for that matter, from a hunter's or fisherman's weekend-long quest for a trophy.

Not that such keyhole glimpses of the natural world don't have a certain value. It's just that, as most of us are only too painfully aware, they are far too artificial, fleeting, and stripped of context. They also tend to be cerebral, more matters of the head, than experiental matters of the heart. Or worse, they are mere diversions — "nature entertainment" — a trivialization that neutralizes nature's power to shape and lend rhythm to the stages of our lives.

Indigenous peoples around the world have long understood that one of the most effective ways to encourage their young (in particular young boys) to form lasting bonds with their natural surroundings has been to culturally sanction, at a critical point in each child's growth, a solitary sojourn in nature. This is a naked time, filled with vivid dreams and visits by animal spirits, during which the familiar world of the child is suddenly turned inside out — a time when spiritual myths are seemingly rendered personal, potent, and emotionally true. For most children of this modern urban age, there is little hope (with a few rare exceptions) of participating in anything resembling the profoundly transformative "curriculum" in environmental education that a young Plains Indian boy, for example, might have received during his vision quest. But a fortunate few among us — wildlife biologists, commercial fishermen, park rangers, for instance — are still granted on occasion the precious gift of prolonged periods of time alone in nature. If we are doubly fortunate, that experience will provide us with moments and memories that forever change the way we see the natural world. And when such internal "re-visions" of the cosmos seem to *sanctify* all of nature, they may bear faint resemblances, however fragmentary, to those insights that traditional native elders sought for their children.

In my own case, despite an early upbringing centered in the Pacific Northwest that offered more than ample opportunity for hiking, camping, and other outdoor experiences, I can recall no single instance of childhood epiphany that gave me a clear sense of my proper place as a human being in the vastness of nature. Not until I reached the age of 25 or so, as a graduate biology student, and found myself "alone" month after month in the company of hordes of wild seals did I experience anything of the sort. And even then, surrounded as I was by steely-eyed fellow scientists, I was usually careful to keep my stirrings of sheer ecstasy to myself. If my article "Birth of a Harbor Seal," published in 1974, does not always reveal such intensity of feeling, at least it conveys some sense of its setting, the murky intertidal waters of Humboldt Bay in northern California, where during the early 1970s I studied harbor seal social behavior. Appearing in *Natural History*, the venerable magazine of the American Museum of Natural History, the article also represented my first proud appearance in print as a bona fide science writer. The second piece, "The Case of the Missing Monk Seal," also published in *Natural History*, combined my professional passion for marine biology with my lifelong passion for adventure-filled travel to faraway places. To research it I spent weeks in Belize and Mexico (after surviving, in the Maya village of Santiago Atitlán with a friend, the infamous Guatemalan earthquake of 1976 that killed more than 20,000 people). I interviewed scores of local Caribbean fishermen and others about reported sightings of a species of tropical seal that most experts feared, with good reason as it turned out, had only recently been hunted and harassed to final extinction. The last article, "Imps of Darkness" — the colorful name that Charles Darwin, sailing aboard the *Beagle*, gave the fierce-looking but docile marine iguanas of the Galápagos Islands — is also in part a tale of historic human abuse of the environment. And it too reflects my nomadic habits and continuing love of the sea.

Harbor Seal Mother and Pup: A Primal Bond

THE PREGNANT COW SEAL is restless. She hitches her swollen body across the mud flat, her stubby fore flippers slapping rhythmically as they pull her along. She pauses for a moment, her eyes glazed and unseeing, then lifts her hind flippers high in the air in obedience to unseen internal forces. In an explosion of brilliant red she gives birth to her pup. The newborn seal is jettisoned headfirst from its mother's womb into a cold world of mud and water; the umbilical cord is severed by the force of the expulsion. At the instant of birth two gulls alight nearby, ready to consume the detritus of parturition.

Stunned and seemingly unaware of each other, mother and pup lie still upon the mud for several seconds. Then the cow approaches her pup and begins to nose it, sniffing deeply of the warm odors by which she will identify her offspring in the weeks to come.

The dark and slender form of the pup also comes to life — first thrashing awkwardly across the mud, then rushing headlong into the chill waters of the bay's rising tide. Within moments the precocial pup is swimming strongly, its mother in close pursuit. As dusk begins to smother the greens and blues of late afternoon, both are lost in the mirror surface of the windless bay.

Such are the first moments of what may be the most enduring social bond in the life of the Pacific harbor seal (*Phoca vitulina richardi*). Following a birth so discreet that few human observers have witnessed it, the cow and her pup embark on a phase of their lives that will bind them in an intimate relationship lasting almost six weeks. It is a bond marked by a continuity that contrasts with the apparent social indifference of adult harbor seals.

The life cycle of the harbor seal is not well understood, and despite the wealth of anatomical and physiological information the animal has yielded in the laboratory, it remains largely an enigma to field biologists.

Groups of Pacific harbor seals emerge from the water to rest on offshore rocks, sandbars, and tidal mud flats in a range that extends from Baja California to the Bering Sea. In some areas along the coast these aggregations are seasonally enlarged by the arrival of pregnant females returning to traditional rookeries to bear their young.

One such rookery is in Humboldt Bay, a broad and shallow body of water that serves as a harbor for the city of Eureka in northern California. With each low tide, the drainage of the bay's waters into the sea exposes a vast, channel-rift plain of mud, which serves as a hauling or terrestrial resting ground for the seals. Hundreds of them emerge from the murky waters during each ebb to "haul out" on the mud flats in widely scattered herds. After several hours, the rising tide covers the bay's muddy bottom, and the seals retreat into the water.

During the first months of the year, the immigration of seals into the southern reaches of the bay is most likely the result of a local centripedal movement of seals from the surrounding coastal hauling areas rather than a linear migration.

Lacking the temporal and spatial integrity of sea lion or fur seal rookeries, the harbor seal herds appear to be almost casual associations, which disperse with each flood tide and regroup, seemingly without plan, on each ebb. Mothers with pups wander freely from herd to herd, often staying on the peripheries, where haul-out space is less competitive. Other seals lie on their sides or bellies in a mass lethargy that only rarely erupts into sporadic, fore-flipper-waving disputes between neighbors. This extremely low level of activity and the difficulty in determining the sexual composition of the herd (harbor seals exhibit little sexual dimorphism) have almost completely obscured the dynamics of harbor seal social life.

In 1973, the births began in Humboldt Bay in late March and peaked in late April, when there were 88 sleek pups and their mothers scattered throughout the colony of 337 seals. For the first days after the brief episode of parturition, mothers and pups face a test of endurance. Although the pup is capable of swimming within a few minutes of birth, it is forced to remain in the water for hours at a time while the hauling grounds are under water. Thus, the selective value of precocity in young harbor seals is apparent.

In the face of such demands on her pup, a mother will frequently carry it piggyback as she swims, becoming a living, mobile hauling ground for her offspring as the two dive and surface in graceful tandem.

Although aquatic nursing has been reported in harbor seals of the British Isles, the nursing activities observed in Humboldt Bay were strictly terrestrial. To make the nursing connection the cow rolls on her side, presenting a pair of retractable nipples situated low on her abdomen. With a facility that improves as the weeks pass, the pup locates a teat and suckles, periodically switching from one nipple to the other. Nursing sessions usually last five to 10 minutes, but may be spaced several hours apart because the rich milk of harbor seals contains as much as 45 percent fat.

Occasionally a pup will show a reluctance to nurse. When this happens, the mother will urge her offspring by waving a raised fore flipper at the recalcitrant youngster (a threat gesture common among adult harbor seals); or she may gently nudge the pup toward her belly with her fore flipper.

The pup's growth over the weeks is testimony to the success of the feeding regimen. Weighing about 25 pounds at birth, a pup may more than double its body weight by the end of the five- to six-week nursing period, metabolizing the fat-laden milk into insulating capsules of blubber.

Looking like tiny, bloated caricatures of adults, suckling harbor seal pups wallow beside their mothers on the hauling grounds. Their eyes are ringed with circles of tears because this species lacks any nasolacrimal duct to drain eye fluids into the nose. This is a feature shared by other marine mammals whose aquatic habits make such a conduit unnecessary. The velvety coats of the pups are mottled silver and black. While still in its mother's uterus, each pup had a pale and silky fetal coat known as lanugo, but this was shed and replaced by the present pelage before birth.

The pups coo softly and sometimes flail their fore flippers or nip playfully at their mothers — often displaying crude but recognizable elements of adult behavior. Unlike sea lions or fur seals, whose pups lead active and playful lives within highly structured, territorial breeding colonies, unweaned harbor seals do not interact with each other or with other adults. Thus, the development of a harbor seal pup takes place in something very close to a social vacuum. Realizing the role that the maternal bond plays in the socialization of some animals, an observer may be tempted to speculate on its effect on harbor seal society. Perhaps such a socially insulative bond plays a part in creating a harbor seal social system in which adults interact only rarely outside of sexual encounters — in what might be called a seal "society of strangers" when compared with more socially gregarious seals, sea lions, and walruses.

As they rest on land, mothers periodically nose their pups, as they have done from the moment of birth. It seems an effective means of identification, for evidence suggests that mothers can distinguish strange pups and will nurse only their own. Even as she swims through the labyrinth of channels that course through the mud flats of low tide, a mother seal continuously monitors her pup's presence by nosing it each time the pair surface from a dive.

In contrast to the almost mute adult, a harbor seal pup is a vocal creature. After a disturbance, the pups' chorus of plaintive cries shatters the otherwise almost total silence of the rookery. Should a storm separate a mother from her pup, the pup's hoarse call will enable the mother to navigate to its side, where she will perform the final olfactory test of maternity.

By the discriminating powers of olfactory, acoustic, and, perhaps, visual communication, harbor seal cows maintain the exclusive bond with their pups. The bond is continuous until the pup is weaned — a mother will not abandon her pup for even brief periods of time unless disturbed. As a result, it is likely that harbor seal mothers either fast or feed sparingly until their pups are weaned. And, in an abstinence imposed by motherhood, the cow will aggressively thwart the ardent approaches of bulls.

Not until after weaning, when she is unencumbered by maternal duties, is the female ready to mate again in an aquatic union — the details of which have eluded naturalists for years. A harbor seal cow is thought to enter estrus a few weeks after her pup is weaned. Although it is not uncommon to see pairs of seals engaging in rolling aquatic interactions at this time — nuzzling, splashing, diving, and sometimes mounting briefly at the water's surface — the sexes can seldom be distinguished.

Aquatic copulation after an estrus-free maternal bond may help to explain the absence of harems or territories in the harbor seal — familiar features of the highly structured societies of otariid, or eared, seals, which have aptly been called "cities of seals." In a seal species in which mothers are not sexually receptive until after their land-based nursing obligations are complete, a male breeding strategy of land partition would not be adaptive. In a species in which estrus occurs during nursing, however, territoriality would provide males with exclusive access to receptive females.

That the former strategy characterizes most phocid or earless seals (including harbor seals), and the latter the otariid or eared seals, suggests that selective pressures may have appeared early in these pinniped lines. Of the *Phocidae*, or true seals, only the elephant seal and the gray seal appear to mate before weaning.

What these hypothetical selective pressures may have been remains speculation, as does much of pinniped evolutionary history itself. If, as some biologists believe, earless seals are descended from an otterlike ancestor that inhabited the inland waterways of Tertiary Asia, predatory pressures may have been crucial. Unlike the shy ancestors of modern eared seals, which formed rookeries on offshore rocks and islands, the earliest earless seals may have been especially vulnerable to land predators. Predation, especially during the reproductive phases of its life cycle, may have selected for a creature that mated in the protective cover of water and provided brief but intensive care for its offspring in a maternal bond undisturbed by the distractions of mating — one not unlike that of the most primitive of true seals, including the harbor seal.

In a society built upon such a primitive breeding plan, harbor seal males would be expected to adopt the more effective strategy of promiscuity. Focusing not on the land resource, which in many sea lions and fur seals, for instance, guarantees access to receptive females, but instead on the mobile females themselves, a harbor seal bull might consort with an individual cow as she approached estrus. After the onset of estrus and an underwater copulation, however, a bull might then move on in search of other potential mates among the postweaning receptive females. Encounters between rival bulls would result in battles and possibly in the formation of bachelor herds of subordinate males. Herds composed, as far as could be determined, entirely of adult males have, in fact, been seen in the Humboldt Bay rookery.

The notion of such a promiscuous system, built upon a primal maternal bond that precludes breeding, lends order to what otherwise seems an anonymous horde of seals. But until more evidence emerges, it remains speculation, and the sex life of this society remains the harbor seal's secret.

1974

The Case of the Missing Monk Seal

In late August 1494, a ship commanded by Christopher Columbus, then on his second voyage to the New World, lay anchored off the island of Alta Vela, just south of Haiti. A shore party disembarked and approached the rocky island. On the beach, they chanced upon a herd of eight "sea wolves," or seals, each about eight feet long, resting peacefully. Their brown pelages were lightly frosted with gray, fading to pale yellow on their stomachs, and hoodlike rolls of fat surrounded their retracted necks. They closely resembled the monk seals that some crew members might have seen in the Mediterranean's more familiar waters. Astonished by the animals' lack of fear, the men crept closer and within a few minutes easily killed all eight.

And so ended, in a prophetically bloody manner, the first recorded encounter between Europeans and the sea mammal now known as the Caribbean monk seal, *Monachus tropicalus*. (Species classified in the genus *Monachus* are the only tropical seals.) Caribbean monk seal herds once thrived along the shores and islands of the Caribbean and the Gulf of Mexico, but harvested and harassed over the past five centuries, their populations have steadily dwindled. By the late eighteenth century a sighting had become a rare event. The last authenticated sighting of a Caribbean monk seal occurred in 1952 on Serranilla Bank in the western Caribbean, and the scientific world has all but conceded the animal's extinction. Ask almost any resident within the monk seal's former range — south from the Bahamas and northern Gulf of Mexico as far as Honduras, east from the Caribbean shores of Mexico and Central America to Jamaica, Cuba, and Haiti — whether he has ever seen a seal. His answer will almost certainly be no; probably he will softly assure you, as if speaking to a foolish child, that seals are creatures of colder seas and currents and have never lived in these tepid waters.

Nevertheless, since 1900 the Caribbean monk seal has been seen at various locations — the Tortuga Islands, Key West, the Texas coast, the waters off Kingston, Jamaica, and at Isla Mujeres near the Yucatán Peninsula. Many of these sightings, including several within the past two decades, remain unconfirmed because a trained scientific observer was not present; other reports are hearsay or may have been sightings of other sea mammals in the same waters — the manatee, *Trichechus manatus*, which appears regularly, or the occasional California sea lion that has escaped captivity. Yet, persistent seal sightings in recent years — infrequent and vague as they are — by fishermen from Yucatán and Belize must, like any unseen animal's spoor, tantalize even the most fatalistic biologist.

The Caribbean monk seal's life history is almost as mysterious as the question of its current existence. Beyond its incredible lethargy and fearlessness on land, and some evidence that single coal-black woolly pups probably have been born in parts of the Caribbean in early December, we know almost nothing of its habits. We do possess some preserved skin and skeleton specimens, which have provided anatomical descriptions of monk seal characteristics — for example, an unusual skull shape, four rather than two mammary teats, and structural modifications of the flippers. But specimens did not reach European museums until the mid-nineteenth century when the monk seal population already was depleted.

Some scientists, in their zeal to learn more about the animal, probably contributed to its decline. In December 1886, a joint Mexican-American expedition to study monk seals spent three days on the Triangle Islands west of the Yucatán Peninsula. Although the expedition gathered undeniably valuable scientific data, it killed at least 40 seals — and preserved only a fraction of them — at a time when the Triangles were perhaps one of the last breeding grounds of the Caribbean monk seal. About the turn of the century, collecting for zoos and aquariums resulted in seal deaths in the field and in captivity.

We do know that there are three geographically distinct species of monk seals classified within the genus *Monachus*. By this century, all three species had shrunk to perilously low numbers, although only the Caribbean monk seal seems to have slipped from scientific view. Its widely separated relatives, the Mediterranean monk seal, *M. monachus*, of the Black Sea, the Mediterranean, and adjacent areas, and the Hawaiian monk seal, *M. schauinslandi*, from waters around Hawaii and Laysan, definitely survive, and may number from 1000 to 5000 and from 500 to 1000 respectively.

Monk seals are peculiar among the Pinnipedia (an order of aquatic carnivores that includes seals, sea lions, and walruses) in their year-round preference for tropical seas. In this habitat, their conventional pinniped adaptations to cold water, such as insulating coats of blubber and high metabolic rate, would seem hindrances. We do not know exactly how monk seals reconcile their body plan with their fondness for warm waters; we can only assume that they have evolved physiological and behavioral strategies to avoid overheating.

Judging by similarities in teeth and skeleton, monk seals are most closely related to seals of the Southern Hemisphere — the Weddell, crabeater, leopard, and Ross seals of the Antarctic. Monk seals probably emerged from a parent stock in the Mediterranean, whence some may have drifted astride the Canary and North Equatorial currents to seed the West Indies with their kind. A few may have penetrated the narrow Isthmus of Panama and eventually crossed the Pacific to the Hawaiian Islands. Reproductively isolated from each other by oceans, these three monk seal populations gradually underwent sufficient independent evolutionary change to warrant their current classification as separate species.

Although Caribbean monk seals must have been hunted previously, they were apparently numerous when the first Europeans arrived. However, from the seventeenth until the late nineteenth century, the seals were relentlessly exploited for the commercially valuable oil produced from their fat. In the early 1700s, a local Bahamian historian indicated the proportions of the rich seal fishery and its potential for destroying its resource: "The Bahamas are filled with seals; sometimes fishers will catch one hundred in a night. They try or melt them and bring off their oil for lamps to these Islands."

Vulnerable on land, like other seals, monk seals suffered a further disadvantage: since seal-hungry predators had not been a part of their tropical environment, the monk seals failed to evolve the innate suspiciousness characteristic of many seals. Disarmed by their own calm dispositions, the sluggish monk seals succumbed when introduced to a new force of natural selection — club-wielding men. And local fishermen have sniped relentlessly at seals, regarding them as competitors for fish, as well as sources of income from blubber, hides, and meat.

Besides these pressures, recent evidence indicates that the ultimate factor in the decline of Caribbean monk seal populations may simply be too many people: seals have been losing their habitat to an expanding human community. In 1973, wildlife biologist Karl W. Kenyon carried out, under the auspices of the Office of Rare and Endangered Species and the U.S. Fish and Wildlife Service, a 4000-mile aerial search for the Caribbean monk seal. Kenyon reported:

> At every island group visited, we found either fishing vessels or shrimp trawlers at anchor, or fishermen and their shacks on shore, or the remains of abandoned fishermen's camps. We looked without success for any indication of the existence of monk seals . . . my conclusion from the 1973 survey is that the Caribbean monk seal has been extinct from the early 1950s. . . . Even had a few old Caribbean monk seals survived to the 1970s, all available evidence leads me to believe that there is no hope that the species can recover. Man has now dominated its environment.

While the survey is necessarily inconclusive, its failure to locate any remnant monk seal colonies in the most promising parts of its former range must dampen hopes for the Caribbean monk seal's survival. But reliable sightings within the past 30 years, plus other unconfirmed reports by fishermen, encourage a continued search. During March 1976, I spent two weeks in Honduras, Belize, and Mexico and ran across several intriguing accounts, along with numerous negative reports, of seal sightings in Caribbean waters. Traveling overland from one coastal fishing town to the next, I questioned local fishermen on the wharfs and in the markets. Many seemed unable to distinguish between manatees, otters, and seals; most, upon studying an illustration I carried, quickly denied the monk seal's existence in the Caribbean. One Belizan conch fisherman insisted that neither he nor his 86-year-old father, in their lifetimes of fishing, had ever seen or heard of seals; my reports to the contrary, he said, were simply lies.

But erratic as they are, reports to the contrary do turn up. A young fisherman from Belize City talked of seeing several very young seals, each weighing about 40 pounds, swimming near offshore reefs between Punta Gorda in southern Belize and Livingston, Guatemala. Even there, he admitted, the seals were a very rare sight, visible only at dusk or dawn.

Another fisherman, a middle-aged man from Stann Creek, Belize, spoke confidently of seal sightings some 20 years before on the beaches of the uninhabited Mexican atoll, Chinchorro Reef, east of Yucatán Peninsula. Here, he said, the seals had often proved a nuisance by becoming entangled in turtle nets.

In Chetumal, Mexico, an elderly and obviously poor fisherman mentioned still other seal sightings at Chinchorro Reef, where he claimed to have seen the animals dive from rocks on which they were basking and disappear into underwater caves. Prodding the old man for further details, I patiently explained the seal's plight. Alert to the seal's rarity, the man leaned forward and asked earnestly, "Could a man sell the meat of such a seal for much money?" I was sharply reminded that concern for the welfare of endangered beasts is a luxury to people with empty bellies.

At the very least, these and similar unauthenticated reports of monk seal sightings underline the difficulty of finally establishing a species' extinction. Local fishermen log far more time in the monk seal's waters than most expeditions can afford, and hence their reports could conceivably constitute a sort of grass-roots monitoring system of a remnant breeding colony. In fact, anyone within the Caribbean monk seal's range who spots a nearly uniform brown seal, without apparent external ears, tinged with gray and fading to pale yellow or yellow-white ventrally, measuring six to nine feet from nose to tail, and weighing perhaps 550 pounds should photograph and report it to local and international conservation authorities.

The case of another supposedly extinct seal, the Juan Fernández fur seal, *Arctocephalus philippii*, of South America, may serve as an inspiration in the search for the Caribbean monk seal. Once common on a cluster of islands some 400 miles off the coast of Chile, the Juan Fernández fur seal also suffered terrible commercial exploitation, beginning in the eighteenth century. By 1900, the species had been all but eliminated from its island range.

While scientists were lamenting the fur seal's disappearance, Chilean fishermen of the Juan Fernández Islands continued to encounter them from time to time in remote fishing grounds. Not until 1958, when a series of documented fur seal sightings began, did the scientific community begin to accept the "rediscovery" of the Juan Fernández fur seal. Reduced to several hundred, and inclined to retreat to isolated sea caves when disturbed, the seals had for a time been lost to scientists, but to sharp-eyed resident fishermen the animals had never completely vanished.

However, the monk seal's critical vulnerability to human disturbance and harassment, especially during the breeding season, and the amount of human activity in the Caribbean weaken the animal's chances of survival. Monk seals, it appears, simply will not breed successfully in the face of continual human disturbance, which poses a serious problem in their management. The case of the missing Caribbean monk seal may be another instance of international concern arriving too late. Had recent proposals for a monk seal breeding sanctuary been put forward and acted upon early in this century, a more optimistic prognosis for the species might be possible.

Although its memory lingers fresh and close, the Caribbean monk seal will probably remain a stranger. And at times, I find myself wondering what dark and fleeting image of species-end might have passed — barely sensed like a frigate bird's shadow — before the eyes of a monk seal as it lay, not long ago, on the sands of some remote Caribbean key.

1977

Imps of Darkness: Marine Iguanas of the Galápagos

BASKING HIGH ABOVE the surf on a boulder of dark volcanic basalt — so close that I could touch its dry parchment skin — sprawls an adult marine iguana. It is a member of a Hood, or Española, Island population located in the southeastern portion of the Galápagos island archipelago, some 600 miles west of the Ecuadoran coast of South America. Perhaps it is a direct descendant of a specimen that Darwin himself observed before scribbling in his notebook during his five-week visit to these islands in 1835 aboard the H.M.S. *Beagle*.

The lizard's three-foot-long body is the dull black of sooty cast iron, reddened with irregular splotches of pigmented scales. As it raises its expressionless head and cocks it to one side, alert to this intrusion, I look more closely for the features that distinguish this species from its terrestrial counterpart on the Galápagos. There is the blunted snout that enables the marine iguana to closely crop the pastures of marine algae on which it feeds, and for which it received its scientific name, *Amblyrhynchus cristatus*, in 1825. Its powerful limbs terminate in elongate digits equipped with long, curved claws that can grip the slippery ocean floor during feeding. The muscular tail is also peculiar — laterally compressed into a swimming organ whose serpentine movements propel the reptile through the sea. Less shaped by the animal's aquatic habits, but no less impressive at close view, is the distinctive dragonlike mail of intricate scales covering its body. An erect, linear fringe of whitened scales extends down the back of the beast's neck, and proceeds down the length of the spine as far as the tail base. On its head rests a bleached helmet of stout conical scales, useful at times in head-to-head pushing matches. Fine-scaled skin hangs like cloth over its throat.

The iguana turns its head quickly to follow a movement with its deep-set crocodilian eyes. Then, frightened, it shoots two jets of liquid from its scaled nostrils and scrambles gracelessly to a nearby rock to join a horde of its fellows.

In his brief visit to this cluster of protruding volcanic mountaintops rising some 7000 to 10,000 feet above the ocean floor, Darwin, as history has amply documented, observed a number of curiosities — ranging from giant tortoises and flightless cormorants to arborescent cacti and tool-using finches. The peculiar habits of the seagoing iguanas of the Galápagos Islands — "imps of darkness," he called them — were certainly among them. He described the species as "most remarkable, because it is the only existing lizard which lives on marine vegetable production" and commented upon its natural history frequently in his descriptions of the islands' strange flora and fauna.

Darwin puzzled over the question of the species' aquatic sojourns, and its precise underwater food. Despite the written reports of the islands' previous visitors that the amphibious lizards "go to sea in herds a-fishing," Darwin set about to establish the facts. His dissections of collected specimens during the voyage established at once that the animals did not feed upon fish: the stomachs contained only the finely minced fragments of green or red algae, with occasional remnants of marine crustaceans. Subsequent observers of the animal have further documented its vegetarian habit, which relies almost entirely on a number of species of benthic and intertidal marine algae that make up the relatively rich marine flora of the Galápagos waters.

In order to forage successfully for food in deeper waters offshore (the marine iguana has been observed feeding in waters as deep as 35 feet), the creature must be capable of diving for a considerable period of time — a fact that one of Darwin's equally curious crew on the *Beagle* effectively demonstrated. The seaman attempted to drown an iguana by sinking it with a weight attached to a line; an hour later he drew up the line only to find that the animal was still alive and active. The *Beagle*'s "experimental results" remain unexceeded to date.

Darwin himself aptly described the marine iguana's aquatic grace, writing: "When in the water this lizard swims with perfect ease and quickness, by a serpentine movement of its body and flattened tail — the legs being motionless and closely collapsed on its sides."

The marine iguana's feeding excursions are generally carried out only once each day and may extend as far as a half mile offshore. The strenuous activities of swimming through waves, diving for prolonged periods, and grasping the rocky sea bottom while tearing at the seaweed growth with rasplike tricuspid teeth are usually followed by a return to rocky resting areas along the shore to bask. Nights are spent sleeping either clinging to the walls in vertical rock crevices or lying beneath a canopy of shoreline vegetation.

The *Beagle*'s visit to the islands did not coincide with the reproductive season of the marine iguana populations, which commences in the first months of the calendar year. Observed Darwin:

> During our visit [October], I saw extremely few small individuals of this species, and none I should think under a year old. From the circumstance it seems probable that the breeding season had not then commenced. I asked several of the inhabitants if they knew where it laid its eggs: they said that they knew nothing of its propagation, although well acquainted with the eggs of the land iguana kind — a fact, considering how very common this lizard is, not a little extraordinary.

It is during the breeding season that the gregarious marine iguana, like other polygamous species, readjusts the age and sex composition of its terrestrial herds. Competition between adult males for territories on the haul-out rocks leads to the formation of female harems in these areas and separate herds of nonbreeding and juvenile animals in other areas.

Male territorial battles consist of highly stylized threat displays: the combatants present their side profiles — inflating throat and body regions to enhance their effect — and engage in bouts of head bobbing until dominance is effectively established. Occasionally interactions escalate to actual head-to-head pushing contests, each animal relying upon its distinctive heavily scaled head plates to take the brunt of his opponent's blows.

Mating takes place on the dominant males' territories after females are courted with a variation of the threatening head-bobbing behavior. Inseminated females then proceed to a nesting area along the shore where they dig shallow burrows in which to lay their white, three-inch-long, soft-shelled eggs — usually two per female. (There have been reports of nest burrows located high above the surf on cliff tops; their relative inaccessibility may account for Darwin's local informants knowing so little of their whereabouts.) Approximately two months later, the first hatchlings dig their way out of these subterranean incubating chambers to face the predatory activities of local snakes, gulls, herons, and hawks.

A STRANGE ANOMALY Darwin observed about the Galápagos marine iguana was that when frightened it will not enter the water. After noticing this unusual behavior, the inquisitive naturalist performed one of his customary field experiments: "I threw one several times as far as I could, into a deep pool left by the retreating tide; but it invariably returned in a direct line to the spot where I stood." Darwin's tentative conclusion, and not an unreasonable hypothesis at the time, was simply that the iguana was instinctively drawn to the safety of the shore, owing to the threat of predatory sharks. However, biological studies have since shown that Darwin's reluctant marine iguanas were not so much fearful of the ocean's predators as they were uncomfortable with its cool temperatures. With a preferred body temperature of somewhere between 95 and 99 degrees Fahrenheit, the cold-blooded iguanas simply found their sudden plunge in the sea too chilly to prolong.

Like other reptiles, the marine iguana is ectothermic — tolerating, within limits, fluctuations in body temperature as the ambient temperatures in its Galápagos habitat range between extremes of about 70 degree Fahrenheit seawater and 120 degree Fahrenheit rock surfaces. However, in contrast to most reptiles, which simply move to cooler or hotter spots, the marine iguana depends almost exclusively upon a repertoire of basking postures to regulate its body temperature.

An iguana will willingly enter ocean waters and can feed actively there, despite the cooling effects of a medium that may be 20 degrees below its preferred range. Back on land, the iguana generally assumes a pose that will accelerate the rise of its depressed body temperature. Typically, the lizard will lie with its flanks at right angles to the sun's rays — maximizing exposure to solar radiation — in a markedly flattened, prostrate position with legs, belly, head, and neck in direct contact with the sun-warmed rock substrate below. (It is difficult to observe such a sprawling, solar-heated beast without ascribing to it a certain sense of blissful comfort, if not blatant tropical lethargy, however anthropomorphic such an attribution may be.)

As its body temperature rises, the iguana will change to another posture, with its head toward the sun so that its body is parallel to the solar rays, which reduces the radiant heating effects. It will also elevate itself by extending its front legs and raising its body above the lava substrate. This reduces heat conduction from the rock, while increasing body heat loss by convection to sea breezes.

Field studies in the Galápagos have shown that these simple methods — as well as other more subtle circulator modifications — are critically important to marine iguana thermoregulation. In fact, an iguana that is physically restrained from assuming its elevated basking posture will soon die from heat exhaustion. Immature marine iguanas are less unorthodox in their thermoregulatory behavior: in addition to postural controls, these smaller-bodied animals will often retreat into the protection of deep lava crevices rather than tolerate the temperature swings of one location.

Perhaps Darwin's most prophetic remark in describing his Galápagos visit is reflected in his astonishment "at the amount of creative force, if such an expression may be used, on these small, barren, and rocky islands; and still more so, at its diverse yet analogous action on points so near each other." Although the geographical variation between island populations was more pronounced to Darwin in his observations of the giant tortoises of the Galápagos, he recognized its more subdued presence in the marine iguana as well.

Current taxonomy, while acknowledging but a single species of marine iguana, does document evidence of racial differentiation between some island populations — suggesting that ocean barriers still exist between the aquatic marine iguana populations. And while the precise number of marine iguana subspecies may number seven or more, such variation is dramatically evident in populations such as those on Hood, or Española, Island. Here, breeding adults often take on a characteristic reddish body coloring as well as distinctive reproductive behaviors.

Like so many animal species on the Galápagos, the marine iguana is remarkably docile toward humans. As Darwin noted, "They do not seem to have any notion of biting; but when frightened they squirt a drop of fluid from each nostril." Sprays of hypersaline water from iguanid salt glands, sadly, have had little effect in deterring human intrusions upon the marine iguana's island habitat.

Since the islands were first visited during the fifteenth century by parties of Incas in sailing fleets of balsa rafts, the Galápagos fauna has been exposed to a relentless stream of humanity. Between the arrival of the first recorded European visitors in the early 1500s and the creation of the Charles Darwin Research Station in 1962 through the cooperation of international scientific interests, Galápagos animal populations were subjected to thoughtless despoliations by pirate crews, sealing expeditions, zoo collecting expeditions, and foreign military operations.

Currently, human disruptions are of a different sort. There are more than 5000 settlers in the archipelago, and they have introduced a trail of exotic plants and animals over the years, and the thousands of foreign tourists visiting each year test the patience of even a Galápagos animal with their photographic pilgrimages.

Because of its amphibious ways, the marine iguana may actually prove less vulnerable to such new pressures than many other Galápagos species. But as one of what Charles Darwin recognized as the "aboriginal creations found nowhere else" but these islands, its future must remain today more uncertain than it appeared a century and a half ago to the keen eyes of the visiting naturalist from the *Beagle*.

1980

Part II

ON NATURE WATCHERS

Introduction

IN THE EARLY 1980s I belatedly began to discover for myself the power of biographical articles, or profiles, to convey a sense not just of scientific ideas but also of the way these could clash with real human lives and values. Instead of just pontificating about abstract scientific concepts and concerns I began to enjoy portraying the earnest efforts of individual scientists trying to reconcile such ideas and issues with their own personal and professional lives.

The first profile is a rather brief sketch of the aging American Nobel laureate biochemist Melvin Calvin. His name is immortalized in the famous Calvin cycle, which he traced in plant photosynthesis — and which every undergraduate biology student painfully recalls having to memorize. At the time, two things in particular intrigued me about Calvin: the workings of his obviously brilliant mind and reports of increasingly public eccentricities on his part during his later years. Was a Nobel laureate by definition simply some sort of biological genius? (I don't think so; nor, it seems, did an amused Calvin.) Was a distinguished scientist as vulnerable as you or I to the risks of becoming a bit too captivated by his own incandescent public image? (I suspect so, and in retrospect I think Calvin's idiosyncrasies provided possible proof of the process.)

I must confess that I came across the life and times of the late

Santiago Ramón y Cajal, the Spanish Nobel laureate and neuro-anatomist and one of the preeminent pioneers of modern brain science, utterly by accident. It so happened that not only was he a distinguished research scientist (working in turn-of-the-century scientifically Third World Spain), but he was also a prolific writer, photographer, and painter — in short, a bona fide national hero to most modern Spaniards. So when I chanced one day upon a library shelf full of books both by and about him, many of them written in Spanish, I was instantly mesmerized by this gifted man who so successfully blended native intellect, artist's eye, and boundless personal passion. And I remain so. Scientists like Cajal are, I think, among the rarest birds in all science — and their kaleidoscopic insights often among its most precious.

"Lights, Camera, Science: The Nature of David Suzuki" is a profile of a geneticist who by chance would later become my co-author and good friend. In Canada, and increasingly elsewhere, Suzuki, the host of the weekly CBC television science series "The Nature of Things," is undoubtedly the best-known scientist in Canada. Suzuki seems to inspire in Canadians a remarkably widespread devotion — but also in some quarters it seems, a modicum of equally heartfelt disdain. This is as much a measure, I think, of the polarized state of our society concerning controversial environmental, ethical, and political issues — on which Suzuki has spoken out for a quarter of a century — as it is of Suzuki the man.

He is perhaps best known to the public for his eloquent warnings about the ongoing global ecological crisis and blind spots in the sacrosanct Western scientific/technological worldview, and his concern for ethics in the practice of science. But lately Suzuki seems to have rendered himself even more visible by speaking out with increasing urgency on topics as diverse as the need to save Canada's old-growth forest ecosystems, the promise of powerful new coalitions between Canadian environmentalists and aboriginal peoples, the plight of Amazonian Indians in the tropical rain forests of Brazil, and the tragedy of relying so exclusively upon a free-enterprise economic system to guide our decisions about the management of temperate forests, coastal fisheries, and other complex biological systems. My profile of Suzuki was written in 1985, and owing to its appearance in what turned out to be the last (and undistributed) edition of the American *Science 86* magazine, it was never published. But it remains, knowledgeable readers have confirmed, a rather revealing look at Suzuki's achievements, values, and motives. In any case, because of the professional and personal relationship that has since materialized between us, it will remain, as written then, my final snapshot of Suzuki.

Finally, in "The Heart's Lonely Hunter: A Profile of Canadian Cardiac Researcher Adolfo de Bold," I describe a fiercely independent, highly innovative Argentine- Canadian heart researcher. His portrait echoes in a way some of the same themes of character, creativity, and solitude that I attempted to explore in my biographical piece on Cajal. I wrote the de Bold article strictly "on assignment" for *Equinox* magazine, knowing absolutely nothing of his work or reputation. Only after long hours of conversation with the beleaguered scientist did I begin to realize that the tale unfolding before me concerned another of science's quiet yet in many ways genuinely heroic personalities. It was, to me, a tale filled with myriad meanings.

The Calvin Cycle: Chronicles of Maverick Berkeley Nobel Laureate Melvin Calvin

AUTUMN SUNLIGHT is streaming in through the windows of Latimer Hall, on the Berkeley campus of the University of California, as I pause to view one of several of the Chemistry Department's icons hanging on the wall. It is a large portrait of one of the department's most illustrious members — Nobel Prize–winning chemist Melvin Calvin.

The face that I see gazing out over the scurrying hordes of undergraduates — a somber, smooth-skinned, self-assured face — offers none of the clues that I am looking for. With dark, narrow-set eyes, taut mouth and a high forehead that merges into gleaming baldness, Calvin looks more like some inscrutable bureaucrat than the flamboyant, often controversial, scientific figure that he is.

Six floors above me, the 71-year-old scientist, still actively carrying out research that has spanned nearly a half century on this campus, waits to discuss with me one of his current interests — plant species that can be commercially grown for their hydrocarbon-rich oils as an alternative source of fuel. It is a natural focus for Calvin, who earned his Nobel Prize in 1961 for a series of elegant studies of the chemical processes by which green plants transform elemental carbon, using sunlight, into more complex molecules of sugar — in a circular biochemical pathway, the Calvin Cycle, that bears his name.

As I begin the long ascent to his office, climbing the graceful white stone staircase that sweeps up toward his aerie across the hall from the Melvin Calvin Laboratory, I recall our first brief meeting. When I called on him only a week before at the arranged hour, I witnessed a volatile side of Calvin's personality that others had already alluded to. Calvin had forgotten about our engagement and was busy setting up a projector for a seminar in his spacious office. When he spotted me at his door, he looked furious. "What do *you* want?" he fumed. My explanations seemed only to make him more livid; he paced about the room angrily, muttering about schedules and lecture obligations before abruptly ushering me to his door with no mention of an alternative date for our meeting.

It was only a minor incident but, in the absence of better explanations, seemed curiously relevant to problems Calvin has been experiencing with the press. Calvin has always been an innovative scientific thinker — freely ranging across boundaries between disciplines (into space biology and the origin of life, the chemistry of cancer and of the human brain, to name a few) — but in recent years he has, it seems, begun to generate a backlash.

Already prone to bold public prognostications on a broad range of scientific topics that excite his interest, Calvin has an energetic personal style that has rubbed some reporters the wrong way. At the mere mention of Calvin's name, for instance, an editor of a leading popular science magazine joked openly about the scientist's reputation for showboating. Worse, a science article in the *Boston Globe*, published in October 1982, mocked Calvin's use of an unproven gene-splicing technique to design a variety of local weed (*Euphorbia lathyris*) that would not only thrive on marginal lands of the western United States, but would also — by adding a single gene from a Brazilian tree — ooze a sap that could be burned directly as diesel fuel.

Not only did the *Globe* reporter pan Calvin's scientific claims unmercifully (among them, the notion that acreage equivalent to the size of Arizona could one day grow sufficient oil-producing crops to meet an estimated 20 percent of U.S. petroleum needs), but he also lampooned Calvin, comparing the excitable chemist to "a short, bald preacher vigorously making a point about sin."

Now, suddenly he is walking beside me, lost in private thoughts. Dressed in gray slacks, turtleneck sweater, and leather jogging shoes, he looks younger than his years. At the sound of my greeting he looks up, startled, then smiles warmly and leads me into his office as if the earlier incident had never taken place.

Within minutes, he is reciting tales of his involvement with oil-producing plants like someone who has been asked to do it many times before. His long narration, in rapid flights of words laced with precise statistical details, touches on the historical antecedents to his oil-gushing species of plants. To bolster his case, Calvin gets up from his massive executive desk and steps over to an antique cabinet across the room, where he removes a number of small glass vials. Then, holding each in turn to the window's light, he proceeds to lecture me — in the pedantic style of a high school science teacher resigned to a world peopled by lesser scientific minds — on the virtues of each.

In one vial is the translucent sap of the Brazilian *Copaifera*, bled from the towering tropical tree (the same one that provided the critical gene in his genetic transplant experiments) in maple-syrup fashion — with yields of up to 40 liters of oil from a mature tree each year, he says.

"Here's the oil just as it comes out of the tree," says Calvin, with an almost paternal pride.

Another vial is filled with oil from a local variety of tree. "That one over there, with the orange-colored seed pods," says Calvin, pointing out the window to the tree-lined circular drive below. The U.S. Department of Agriculture is showing interest in that species, he says, because it not only yields a useful oil, similar to turpentine, but also will grow in areas of the Southeast where tobacco farmers are suffering economic hardships. "If the USDA would pick that idea up, it would be great. Not only would it eliminate tobacco — a carcinogen — but it would also produce oil."

WHEN QUESTIONED about his Nobel Prize–winning work on photosynthesis and the events that led up to it, Calvin grows more relaxed; he reminisces fondly about his years in Manchester, England, in the mid-1930s, as a post-doctorate fellow under a mentor — the Hungarian Michael Polanyi — who set the course of Calvin's scientific career. Polanyi, like Calvin, exhibited an interdisciplinary bent; at various stages of his life he worked as a surgeon, a chemist, an economist, a philosopher. He also inspired Calvin to explore the inner workings of a group of chemicals — called chelates — that have a pivotal role in the lives of both animals and plants (among them the hemoglobin molecule of blood and the chlorophyll molecule of green plants, both constructed of highly reactive metal atoms nestled into an organic lattice of carbons).

It was his expertise with such molecules that drew Calvin into the inner circles of nuclear weapons research at Berkeley under Ernest O. Lawrence. It was also this knowledge that prepared him, at war's end, to take advantage of their unmatched supplies ("we had a virtual world monopoly," recalls Calvin) of radioactive isotopes like carbon 14 to study the metabolic processes of living things.

With a smile, Calvin recalls a conversation with Lawrence, only a few days after the end of the Second World War. Lawrence approached him with the suggestion that he consider using the project's stocks of radioisotopes for his own biological research.

"Don't worry about the money," said Lawrence.

"Great days!" replied the ecstatic 34-year-old Calvin.

More subdued now, Calvin relives those early years before the Nobel Prize thrust him, like some other illustrious colleagues, into a public spotlight he may have been ill equipped to cope with. His memory is vivid, his intellectual drive is evident, his personal triumphs are described with humility as he remembers his first attempts to inject minute doses of carbon 14 (in the guise of atmospheric carbon dioxide) into the whirling metabolic wheels of living plant cells in order to track its radioactive spoor and illuminate photosynthesis.

He even refers to the critical sequence of events in his life — like meeting Polanyi, joining the Manhattan Project with its precious carbon 14 and getting a boost from Lawrence, for instance — as an unlikely "concatenation of events," a fortunate interlocking chain of incidents (mirroring, perhaps, the very cycle of biochemical processes he would study) that all but assured him of scientific success.

In the years following Calvin's Nobel award, there was not only an outpouring of honors (appointments to presidential science advisory committees and NASA boards, for instance), but plenty of media attention as well. Here was Calvin, his words captured in newsprint, announcing in a public lecture that there must be life elsewhere in the universe; Calvin editorializing in *Harper's* on the best strategies for cancer research; Calvin and his wife, Gene Elle, co-authoring a plea in *The New York Times* for "large-scale commercial development" of existing hydrocarbon-producing plant species as an alternative fuel source.

But at some point the Nobel laureate's love affair with the media began to sour. "Melvin Calvin, the visionary chemist who gave us gasoline trees, now predicts" begins one 1977 newspaper story. "From the man and the country that brought us the petroleum plant comes now the diesel tree" opens another, wearily.

How does Calvin feel about such public needling?

His initial reply is a shrug. "The media *do* that, and I don't like it," he says with a nervous laugh. "But I figure they'll live and learn." His laughter is tense, like a string of firecrackers exploding, and is tinged with bitterness. "Yes, they're *all* critical now," he murmurs.

There is a temptation to simply lump Melvin Calvin, on such matters, with assorted other maverick Nobel laureates — the vocal scientist-celebrities, in the mold of Linus Pauling perhaps, who seem to thrive on the mix of awe and abuse that the American public heaps upon them. But Calvin positively bristles at the comparison. Without mentioning names, he expresses outrage at the behavior of some of his Stockholm-honored scientific colleagues who "think they are gods." "I hope I don't ever get to be that way," he says earnestly.

He points to his record of staying out of political discussions, confining his soapbox statements to science. "I've avoided political stands in every case," he explains. "I refuse to sign petitions; that's my private business. I'm not campaigning, so I just don't see any reason for all that."

"Should a gifted scientist — one who has succeeded in untangling scientific puzzles that have baffled most of us mortals — be looked to for wisdom on subjects outside his area of expertise?" I ask.

"No, I don't think so," answers Calvin, after some thought.

"And," I ask, "what about your *mind* — do *you* consider yourself in any way better equipped than most people to brainstorm, to solve scientific problems?" It is a brazen question, but one I cannot refrain from asking.

The question hangs awkwardly in the air for several seconds. Then, suddenly, it is lost in peals of infectious laughter, as Calvin's cackle rises into an uncontrolled falsetto that brings tears to his eyes.

"Oh, golly," he moans, gasping for breath. Then, in a final statement on the matter: "Oh, golly!"

Apparently not.

AND WHAT DOES the future hold for Melvin Calvin? There is his demanding lecture circuit, he says, fingering an appointment book packed with speaking engagements throughout the U.S. and around the world. And he will press on with his scientific research "as long as I have graduate students."

Lately Calvin and his co-workers have been developing new schemes to mimic photosynthesis in order to generate fuels. Such "artificial chloroplasts" have already produced bubbles of hydrogen gas from sunlight, but Calvin is hoping to devise a way to funnel solar energy directly into long-chained carbon molecules, as plants themselves do. He points up at his latest experimental design — a U-shaped tube of glass that relies on dyes, suspended in liquid, to attempt to do what comes naturally to plant cells.

"That," says Calvin, "is where the ultimate future lies."

When Calvin speaks of the future, often in tones so soft that they are barely audible, it becomes apparent that it is his age, more than his critics, that disturbs him most. He will talk excitedly about what he would like to try next in some field of research, then, sadly, append the phrase, "or someone else will eventually do it," as if he senses that some of the questions he cares most about might well outlive him.

As I prepare to depart, Calvin is feverishly flipping through the pages of a university catalog, trying to locate for me his name in a list of several so-called University Professors — distinguished faculty members whose appointments encompass the entire University of California system. When he finally locates it, he is visibly upset; someone has placed suddenly after "Melvin Calvin" "emeritus" — an honorary title reserved for highly respected academic retirees.

"I don't like this," he fumes, pointing to the offending word. "They ought to scratch that out. It sounds like I'm over the hill."

At this poignant moment, his forlorn expression bears a frank admission of mortality — one that is seldom apparent in the showmanship of his public persona. As with all of us, it seems, Melvin Calvin's essential challenge is the passage of *time* — the relentless turning of his own life's cycle.

1983

Painter of Neurons: The Artistry of Spanish Nobel Laureate Neuroanatomist Santiago Ramón y Cajal

DEFTLY DEPLOYING lantern slides and striking, hand-drawn charts, neuroanatomist Santiago Ramón y Cajal lectured the distinguished members of the Royal Society of London on the anatomy of the nervous system. A friend in the audience that evening a century ago, British neurophysiologist Charles Scott Sherrington, described Cajal's performance in a memoir:

> He treated the microscopic scene as though it were alive and were inhabited by beings which felt and did and hoped and tried even as we do. . . . A nerve-cell by its emergent fibre "groped to find another"! . . . He would envisage the sperm-cells as activated by a sort of passionate urge in their rivalry for penetration into the ovum-cell. Listening to him I asked myself how far this capacity for anthropomorphizing might not contribute to his success as an investigator.

Cajal's fervor, combined with his meticulous pen-and-ink drawings, enabled Spain's most famous scientist to convey his considerable insights into the workings of the brain. His imagination and passion moved him to advance and defend a theory that many members of Europe's scientific establishment first ignored and then tried to reject. His later work based on that "neuron doctrine" would earn him the 1906 Nobel Prize in physiology and medicine.

Two years before his death in 1934 at the age of 82, Cajal scribbled a final defense of the idea. Its title, *¿Neuronismo o reticularismo?* ("Neuron theory or reticular theory?") addressed a debate still kept alive by a stubborn few scientists: Was the nervous system built of individual nerve cells separated by tiny gaps, as Cajal insisted? Or was it a continuous web, or reticulum, of protoplasm through which the brain's minute signals flow like water along labyrinthine channels? Even a residue of resistance to the neuron doctrine drove the prickly scientist — by then feeble, hard of hearing, and suffering from arteriosclerosis of the brain — to distraction.

Until Cajal, most biologists believed that the neurons of the brain and spinal cord were joined in a gapless web. The problem was that these reticularists, as they were called, could not show conclusively that the axons and the dendrites of neighboring cells were actually fused. Cajal referred to this hypothetical neural network derisively as an "unfathomable physiological sea."

To prove the neuron doctrine, Cajal harnessed a cell stain developed by Italian neuroanatomist Camillo Golgi. Nobel laureate David H. Hubel, a Harvard neurobiologist, calls Cajal's subsequent forays into the previously undescribed cellular universe of the central nervous system "a lifetime of stupendous creativity." His magnum opus — *Structure of the Nervous System of Man and Vertebrates*, published in Spanish in 1904 — is "the most important single work in neurobiology," Hubel says.

Cajal rose to prominence in a country far from the center of the scientific world. Spain's laboratories were pathetically equipped, its libraries sparse, its scientists long accustomed to deferring to their northern European counterparts. To be heard, a Spanish scientist often had to offer discoveries too astonishing to be ignored — and then describe them in a tongue more acceptable than his own. Even Cajal's surname, Ramón y Cajal — a traditional union of his father's and mother's last names — perplexed foreign scientists; to most of them he became famous simply as Cajal.

Born in 1852 into the poverty of Petilla de Aragon, a village of modest tile-roofed houses in the Pyrénées of northern Spain, Cajal was the eldest son of a disciplinarian father who had left farming to learn the surgical and bloodletting skills of a barber-surgeon and, later, physician. The boy had no interest in science; he was a truant and mischief-maker occasionally in trouble with the local police for his pranks. His talent for art, on the other hand, was precocious, but his father could not bear his son's "sinful amusement." Finally Justo Ramón Casasús, his patience spent, dragged the boy, bearing a sketch of the apostle St. James, before the critical eye of an itinerant house painter. "What a daub!" sniffed the painter. "Neither is this an Apostle, nor has the figure proportions, nor are the draperies right — nor will the child ever be an artist." Triumphant, Cajal's father denied the boy all access to art supplies. He yearned for his son, as Cajal wrote in his autobiography, to "renounce my madness over drawing and prepare myself to follow a medical career."

In 1869, Cajal gave in, enrolling in the University of Zaragoza medical school, which was open, like other medical schools of the time, to any youth with a minimal academic background. To keep an eye on his son, the senior Cajal uprooted his family and took a temporary post as an instructor in anatomy. As Cajal helped his father perform autopsies at the hospital at Santa Engracia, he recorded their observations with detailed anatomical watercolor paintings. The father was delighted that his son was employing his talents more acceptably and encouraged him. The younger Cajal enthusiastically adapted his skills into a lifelong research tool.

Cajal earned his medical license four years later. During that entire time, he "had scarcely even seen a microscope," says Canadian neurobiologist and medical historian William C. Gibson, who studied in Charles Sherrington's Oxford laboratory. That was not so unusual in Spanish medical schools of the day. Many professors, Cajal wrote in his memoirs, "despised the microscope" as "prejudicial to the progress of Biology." These teachers, said Cajal, believed that "the marvelous descriptions of cells and of invisible parasites were pure fantasy."

Drafted into the Spanish army upon graduation, Cajal took the opportunity to transfer into the army's medical service as an assistant physician with officer's rank and was sent in 1874 to a remote Cuban jungle. He languished there for a year, disheartened by the oppressive tropical climate and by the corruption he saw all around him. After a near-fatal bout of malaria, he secured a medical discharge. In 1875 he returned to the University of Zaragoza.

Two years later Cajal received an academic doctorate in medicine. But the malaria that nearly killed him in Cuba now weakened him and almost ended his career. He contracted tuberculosis and was forced into a prolonged convalescence. It drove Cajal, given to emotional extremes by nature, to deep despair. At one point, according to his memoirs, he climbed a "lofty crest," where, lying on a boulder recovering his breath, he pictured a solitary death: " . . . letting myself die with my face to the stars, far from men and with no other witnesses than the eagles. . . ." He later attributed his recovery, long months later, to his walks through the forest, burdened with photographic equipment; a diet of meat and milk; and the serene atmosphere of the monastery where he convalesced.

Finally recovered, he found academic positions scarce and often political; it took Cajal seven years to gain a professorship at the University of Valencia, where he quickly became known on campus as approachable but blunt, excessively serious, and obsessed with lab work. He was no foe of the microscope. It is said he once sat at his microscope for 20 straight hours peering at a white blood cell as it oozed through a capillary wall.

Cajal's sudden rise to fame, as so often happens, had its beginnings in luck. In 1887, the young professor with the slightly rumpled appearance traveled to Madrid on business. There he visited "those in the capital who cultivated microscopic studies." One happened to be Luis Simarro, a psychiatrist just returned from Paris, where he had been shown slides stained with the still obscure Golgi chrome-silver technique. Cajal, astonished, peered at a slice of brain tissue impregnated with the stain: against a field of yellow, he saw a sepia constellation of delicate spiderlike shapes — elements of nerve tissue — with radiating processes as fine as thread.

Like other histologists, Cajal had found that conventional stains left axons and dendrites largely invisible, coloring only the cell's nucleus. Golgi soaked the tissue first in potassium dichromate and then in silver nitrate, staining the entire length of neurons whose axons were unshielded by a fatty myelin sheath. Moreover, it thinned the dense jungle of intertwined cells by mysteriously staining perhaps one out of a hundred neurons. But it could take days of soaking for the stain to penetrate. And perhaps because of subtle differences between individual neurons, its performance, at least at first, was so erratic that Golgi finally abandoned it. Cajal set out to refine Golgi's stain, and by 1888 he had already made a major breakthrough: using bird and mammal cerebellum, he confirmed that nerve elements terminate in arrays that can be distinguished from one another and, more important, that do not fuse with neighboring cells. He showed that certain nerve cells, for instance, send long, wandering axon fibers toward the dendrites of adjacent neurons called Purkinje cells, wrapping around them "like ivy or lianas to the trunks of trees" but never uniting with them. Keenly aware of the import of his findings and of the territories left to conquer, Cajal plunged into further applications of the improved Golgi stain, working, he said in his memoirs, "no longer merely with earnestness, but with fury."

The key to his success with the stain lay in his fortunate choice of embryonic and newborn nerve tissue rather than adult tissue. The axons of embryonic neurons, unlike those in adult tissue, are not surrounded by insulating sheaths of myelin, so they absorb stain more readily. "Since the full-grown forest turns out to be impenetrable and indefinable," wrote Cajal years later, "why not revert to the study of young wood?"

Cajal was so anxious to announce his discoveries that in May 1888, using his family's meager savings, he established the *Triquarterly Journal of Normal and Pathological Histology*, ran off 60 copies — only three issues ever appeared — and sent prestigious reviewers complimentary copies. Sherrington would later recall that the ambitious Cajal naively sent a few to some members of European royalty, who must have been baffled by such esoteric mail, if they looked at it at all.

But the major journals ignored Cajal's work. Indignant at the slight, he turned to translating. All the leading anatomists were German while Cajal knew only French, learned under pressure from his father. So he laboriously translated his Spanish manuscripts into French to submit to leading German journals. And he supplemented them with exquisite portraits of neurons he had observed.

Looking like fine etchings of some subterranean realm of tangled roots, the drawings were technical innovation as much as art.

Although the technology of photography was exploding during these years and Cajal was himself an enthusiastic photographer, taking pictures through a microscope was still at a primitive stage of development. To capture the details of a nerve fiber's undulating path through thick tissue sections, Cajal would have had to sandwich together multiple shots, because the lens could not keep the image in sharp focus throughout the entire section. By mentally creating a single, sharply focused neuron from many microscopic views, Cajal the artist communicated visions of neurons that no photographer could possibly equal. "His artistic abilities were crucial," says Gibson. "It was those illustrations that sold the thing."

In 1889, Cajal ventured to a meeting of the German Anatomical Society in Berlin to demonstrate his Golgi-stained slides, a rather bold undertaking for an unheralded scientist from Spain. Apparently he created quite a stir and in the process managed to convert Rudolf von Kölliker, a well-known Swiss histologist and confirmed reticularist, to his point of view. Articles in von Kölliker's journal and others soon confirmed Cajal's findings, abruptly thrusting the Spaniard into the international limelight.

On his way home Cajal stopped in Pavia, Italy, hoping to visit Camillo Golgi and respectfully acknowledge the Italian's contribution to his own work. Golgi, away in Rome carrying out his duties as a senator, was unavailable. It portended the rift, never to be bridged, between the two fiercely competitive scientists. At one point, after Cajal reported the existence of certain nerve fibers in the spinal cord, Golgi announced that he had done so 10 years earlier. And he had, but in a particularly obscure journal. Later Cajal remarked snippily that *he* reported a particularly unnoteworthy cellular structure a year or two before Golgi published a report and got credit for the discovery.

Once back in his laboratory, Cajal responded to his new fame by quickly focusing the Golgi stain, and generations of new stains that followed it, upon other neuronal landscapes — the retina, the spinal cord, the olfactory bulb, and others. If, as Cajal once claimed, "the stature of a scholar is the sum of the original facts which he has discovered," the next years of Cajal's microscopic mapping of the nervous system firmly established his place in neuroanatomy. By tracing the retina's "hard wiring" in many species, for example, Cajal not only identified separate pathways exiting from rods and cones but created an elaborate classification of neurons based on their different shapes.

Cajal devised innovative methods of staining each new region of the nervous system as the need arose. He modified a methylene blue stain to reveal fresh details of the cerebral cortex (using warm corpses supplied cooperatively, if questionably, by Madrid institutions). His studies of the brain's anatomy and circuitry unveiled new varieties of neurons and helped map the areas where, scientists later learned, epileptic seizures often can arise. By showing how the cerebral cortex contains varying densities and kinds of neurons, Cajal provided early evidence of the brain's different functional areas.

His reduced silver nitrate stain, which appeared in 1903, revealed the full outlines of neurons but left them transparent, so that internal details such as delicate neurofibrils could be seen. A gold stain, developed in 1913, colored the unexplored glia cells of the brain, offering later generations of neuroanatomists insights into areas of the brain where tumors develop. Like a camera's tinted filter, each new stain offered a subtly different view of the nervous system, and through the decades of Cajal's research, each of them repeatedly reaffirmed the neuron doctrine.

Camillo Golgi remained a leading die-hard reticularist, in Cajal's judgment "hermetically sealed to all innovation." The two rivals found themselves sharing the stage at the 1906 Nobel Prize ceremonies in Stockholm. Golgi, honored for discovering specific types of neurons and cell structures as well as for his stain, chose to lecture on the reticular doctrine he still found tenable. Cajal, who had been trembling with frustration at the Italian with the handlebar moustache, nonetheless confined his own talk to a review of the evidence supporting the neuron doctrine.

Even as a globally renowned neuroanatomist, Cajal infrequently traveled abroad and remained somewhat provincial. In London in 1894 to address the Royal Society of London, he was entranced by the sight of rush-hour traffic pouring across London Bridge, according to an account by Charles Sherrington, his host. And on a shopping trip, Cajal tried to bargain with clerks as if haggling in a *mercado* in Madrid.

A national hero in Spain after receiving the Nobel Prize, Cajal remained almost obsessed with the fear that the publications pouring out of the Instituto Cajal, established in 1920 and named in his honor, were ignored. "The immense majority of the biologists of today are ignorant of the language of Cervantes," Cajal complained bitterly. "It is not surprising then that in consulting the most recent neurological works, we find with sorrow that two-thirds of the modern contributions from Spain are absolutely unknown."

If Cajal had suffered the anonymity of his fellow Spanish scientists, the emerging field of neurobiology would have lost a scientist who would not separate the world of the microscope from the larger one around him. Cajal deeply believed that the beauty of a single neuron lay both in its minute detail and in its role as part of a grand design. All things, he wrote, should be viewed "from afar so that we do not lose their total enchantment."

1986

Lights, Camera, Science: The Nature of David Suzuki

DAVID TAKAYOSHI SUZUKI's sprawling home in Vancouver looks out over a shimmering city skyline, a distant wall of forested mountains, and choppy Pacific seas that stretch westward toward the Japanese fishing village his grandparents left behind to seek their fortunes in Canada. On one wall of his living room, surrounded by plants, Oriental vases, and rice-paper curtains that flap in the breeze, hangs a framed cartoon of the 51-year-old Japanese-Canadian geneticist. A gift from friends, it depicts the bearded, curly-haired Suzuki dribbling a basketball-sized world. Beneath it are scrawled the words, "Long live the hottest show on earth!"

The show is "The Nature of Things," a weekly Canadian Broadcasting Corporation TV series on science and its impact on our lives. It has aired in Canada in one form or another for 26 years and is currently being seen in more than 30 countries, from Japan and Malaysia to Russia and the United States. During the decade since Suzuki became its regular host and occasional producer, it has helped boost him from a promising young fruit-fly geneticist and civil rights activist to the status of Canada's preeminent popularizer of science.

In fact, it has helped to make him Canada's most "visible scientist," to use the lexicon of Rae Goodell's 1975 book *The Visible*

Scientists. Based on her doctoral studies in communications at Stanford, Goodell's book examines the careers of scientists like Linus Pauling, Margaret Mead, Paul Ehrlich, and Carl Sagan — survivors, she says, of a modern McLuhanesque media world who have earned their popularity less from scientific discoveries or leadership in the scientific community than from their maverick behavior. They are articulate. They have colorful and engaging personalities, media savvy, impeccable scientific credentials. And whatever the content of their diverse messages, it tends to be timely, provocative, and presented directly to the public, often despite the best counsel of more traditional colleagues.

It is such visible scientists who are teaching many people much of what they know of science today. The science they choose to present — and the way they choose to present it — have a significant effect on how the public views science. This may be considerable power in the hands of a small group of the scientific élite, but Goodell thinks they serve a valuable function. "Visible scientists," she say, "are a major benefit to the public understanding of science. The problem is that there are just too few of them to offer us the full spectrum of viewpoints that we need."

The spectrum so far includes such figures as a breathless David Attenborough leading us on ecstatic nature hikes and a courtly Carl Sagan journeying through a multi-billion-starred universe. It could even be said to include the coolly objective, anonymous narrator who presents "Nova"'s elegant weekly treatises.

But it has not, up to now, included scientists inclined to explore the undesirable long-term consequences of scientific endeavors or to genuinely challenge their colleagues. In these terms, David Suzuki seems to occupy a new media niche. Though not familiar to most Americans, Suzuki has become, through "The Nature of Things," a household name in Canada. In addition, he writes a newspaper column on science, technology, and the environment and he anchors a science news report called "Discovery" on rock-and-roll radio stations and even narrates science records for kids. In February 1985 he hosted "A Planet for the Taking," an iconoclastic CBC science miniseries on Western man's relationship with nature that captures better than any previous project his willingness to publicly discuss ethical issues surrounding science.

As any credible visible scientist must, Suzuki addresses such issues from a base of solid scientific credentials. Though he now holds only a part-time faculty position that requires no teaching or research, Suzuki was once one of Canada's most promising young geneticists. Between 1969 and 1972 he held the prestigious Steacie Award as the outstanding research scientist in Canada under 35, and by 33 he was a full professor at the University of British Columbia.

Suzuki's most important work in genetics, carried out in the late 1960s and early 1970s, involved the isolation of so-called temperature-sensitive (ts) mutant traits in *Drosophila*, a class of fruit flies. The fickle nature of such traits can usually be traced to a single amino acid error in a protein. At one temperature the protein's architecture is intact, and the fly appears normal. At another, the faulty amino acid causes a local collapse, leading to the mutant trait's appearance.

At the time, microbes with ts mutant traits were already well studied. But the mutations had not been systematically sifted from the genes of *Drosophila*. Suzuki and his co-workers at UBC came up with a number of strategies for isolating them. One was designed to screen for ts mutations affecting behavior, such as flying ability. Suzuki's team first fed flies a chemical mutagen to induce new mutations. Then they screened the progeny of these flies for the desired trait by placing them in a warm box. The box was rotated so any flies rendered suddenly flightless by the heat tumbled helplessly onto a ledge. These were collected and transferred to a container at room temperature to see which had a reversible paralysis; those that regained flying ability were bred, and the progeny were tested to confirm the genetic basis of the paralysis.

After months of screening hundreds of thousands of flies, Suzuki and his co-workers were rewarded with precisely the sort of genetic freak they had hoped for — a fly that was immobilized at warm temperatures and entirely restored at cooler ones. They value of such a ts mutant lay in the fact that the trait's expression could be controlled by temperature. So a dominant gene, for instance, that normally killed a fly could be preserved for further experiments simply by placing the fly in "cold storage" to deactivate the gene. Or by using temperature as a probe, one could pinpoint the time during larval development when the product of a particular ts mutation was expressed. Over the years, Suzuki's lab gained wide recognition for developing a menagerie of ts mutants — each with a distinctive characteristic. "I became Mr. Temperature-Sensitivity," says Suzuki with a laugh.

In the sanctuary of his fourth-floor UBC office, Suzuki, wearing a T-shirt and cut-offs, leans back in his chair, his bare feet resting on a cluttered desk beside a stack of biographies of famous Canadian scientists and a lap-top microcomputer he uses for writing. The walls are papered with yellowed press clippings, photos, and other memorabilia, chronicling more than two decades as a scientist in the public spotlight. Above him, like a Damoclean sword, dangles a huge white macrame hanging, shaped like an unravelling DNA double helix.

Suzuki praises "The Ascent of Man," "Cosmos," "Nature," and other pioneering science series as "magnificent celebrations of science." But he thinks that TV should begin to offer viewers much more. "My own feeling," he says, gesturing emphatically, "is that the time for showing how great science is on TV is past."

The time he spent working on "Planet," he says, along with the experience of writing a regular newspaper column, has transformed many of his views. He has now emerged as a gadfly of science, speaking out on topics from onshore oil-drilling policies and sexism in science to the failures of science education and the folly of designing software for futuristic Star Wars weaponry.

Despite the diversity of issues, most of his public stands share a few common themes. One is the need for scientists to admit to society that while they can quickly identify the benefits of a new technology, they are seldom capable of predicting its subtle, long-term social and environmental costs. He cites the side effects of birth control pills, the impact of pesticides on eggshell strength, and nuclear fallout as proof that the most important costs of new technologies tend to be hidden from scientists for decades, appearing only after the planet has been used as a testing ground. Scientists, he suggests, are often reluctant to publicly admit such ignorance.

Another recurrent theme is his long-standing belief that scientists have a moral obligation to help bridge the distance between scientist and lay populations. The challenge, as Suzuki sees it, is not simply to demystify science but to empower individuals to respond to any threats science poses to their liberties — a logical extension of his long-standing civil rights interests. "Today," he writes, "our educational institutions fail to educate. Instead they are turning out two increasingly estranged groups, each of them extremely ignorant about the other. One speaks the jargon of scientific terminology, has little grounding in literature, philosophy, history, or religion and yet will become scientists, engineers, and doctors who will apply revolutionary tools to change us and the environment. The other group comprises the majority of society and is alienated from science, believes science is an esoteric activity of the mathematically gifted, and doesn't feel science has much relevance in daily life." The result, he says, is that "most people are powerless to understand or deal with the most pervasive forces shaping their lives."

If such statements annoy some scientists, others have earned Suzuki the label of scientific heretic — for instance, that "the primary motivation underlying the application of scientific knowledge today is destruction and profit." That the bombing of Hiroshima marked "the sanctification of an alliance between science and the military." Or that neutron bombs, binary nerve gases, and biological weapons arise not from military minds but from "the imaginative minds of scientists . . . who are often our colleagues and our students." Because of such statements, some of Suzuki's opponents have written him off, though seldom publicly, as an egocentric media showman and an erratic spokesman for science.

THESE DAYS SUZUKI'S influence is great enough that the criticisms are mostly muffled. Nevertheless, more than one colleague has mentioned that by abandoning genetic research so early, Suzuki "lost his commitment to science." And others have even suggested that because Suzuki devotes the bulk of his time to broadcasting rather than research, he can hardly be classified as a scientist — let alone a visible one.

Suzuki is undaunted — even defiant — in the face of such criticism. In one recent newspaper column, titled "Here is why I continue to cry out against my colleagues," he demonstrates his combative style in dealing with colleagues he sees as "members of a privileged community whose priority above all is to maintain those privileges." Any scientist today, he writes, "who does not feel a need to look at the social and historical context within which scientific 'truths' are obtained and applied runs the risk of repeating past mistakes."

Most scientists find it easy to praise Suzuki's part in "The Nature of Things," a show generally more inclined, with some memorable exceptions, to explore the physics of pitching a curve ball or the physiology of fitness than any hotly controversial issue in science. But with the broadcast of CBC's "Planet" series in the spring of 1985, Suzuki came under heavier fire. Some of the most radical views expressed in "Planet," interestingly enough, originated not with Suzuki but with John Livingston, a former executive producer of "The Nature of Things" and a well-known professor of environmental science at York University. But Suzuki was deeply involved in planning the series, is credited with writing the scripts for two of the eight shows, and never utters a word on camera, he insists, that he "cannot live with" as a scientist.

In the first show, *Human Nature*, host Suzuki raises sweeping questions about the capacity of our Western worldview to satisfy our most basic emotional needs. We human beings have taken control of the planet, he says, and have come to see everything on Earth as a means to our ends. "There is no doubt that through science and technology what we have accomplished is astonishing. But in the worldview of a technological society, we have lost sight of our place in nature, our sense of belonging. Can we find a way to shift our perceptions, to find a view that will encompass both our technical achievements and our ties with our planetary home?"

Later, in *Mythmakers*, he contrasts that scientific worldview with myths of Kalahari desert tribesmen, Hindu villagers, and selected ancient cultures that have contributed to our current worldview, characterized by what he calls the modern mythology of the clockwork universe. This mythology, he says, has given us the tools "to achieve tremendous power over the physical world but by its very success has unleashed a monster with a momentum of its own." He concludes with his hope that "the revelations of science can be viewed in the light of much older beliefs to give us new insights, new mythologies that will reconnect us to the living planet that gave us birth."

The most provocative episode may be *Who Needs Nature*, written by Livingston. It explores our society's willingness to look on animals as curiosities, as entertainers, as victims, as commodities, as companions. In the process, it takes pokes at cowboys, big-game hunters, pet owners, matadors, and even the kindly figure of Walt Disney himself, whose caricatures of animals depict them as "retarded humans."

In an ecstatic Canadian review headlined "David Suzuki saves the world," TV critic Joyce Nelson praised "Planet" as "a fundamental challenge to the dominant worldview of North American society" and "the most profound and radical programming to come through the airwaves in years." Many Canadian viewers agreed; letters to CBC called Suzuki a "member of the twentieth century's chosen priesthood" who "can see where we are headed." Angry letters came too — from creationists, anti-abortionists, and Canadian cattle farmers, among others. Some called Suzuki single-minded, polemical, and self-righteous. One caustically labeled the show "sensational cheesecake for the left-wing righteous and elect."

Such negative remarks from outspoken special-interest groups could perhaps easily be discounted; the pointed criticisms of some other journalists and scientists, however, could not. Stephen Strauss, a science writer for *The Globe and Mail*, Canada's self-proclaimed national newspaper, strongly lampooned the message of "Planet" as a blatant "argument for not knowing." In an open letter to Suzuki, he chastised him for what he perceived as the utter failure of the series to find a balance between its romantic pleas to "let nature be" and the deep-seated human "drive to know and to apply that knowledge."

One scientist who vocally opposed "Planet"'s perspectives was Louis Siminovitch, a prominent Toronto geneticist who has known Suzuki professionally for many years. As a member of the CBC science advisory board that previewed parts of the series before it was broadcast, he was appalled at what he saw as its "decidedly negative attitude toward science" and its failure to "offer practical solutions" to the problems it posed. He fought its release because he felt its net effect would be to leave viewers with a "bad impression" of science and scientists. And he indignantly questioned Suzuki's claim that science itself was the culprit. "Science's imperfections," says Siminovitch, "are a reflection of society's imperfections" — not the other way around.

Suzuki believes that such criticisms arise in part from a fundamental misunderstanding of the goals of the series. "It was not meant to be the definitive scientific statement on the subject," he says. "It was designed as a series of personal essays. Some of the programs are as opinionated and controversial as they could possibly be. But that was their purpose — to try to get people to reexamine their own values." He sees "Planet"'s high ratings (it drew an average audience of 1.77 million — substantial by Canadian standards) as evidence that most science programming tends to be far too patronizing. Such figures "totally put the lie to the general assumption that the TV audience is a bunch of morons," he says. "With 'Planet' you have a series that has emotional impact but a lot of content as well. And the audience has hung in there, not just for one show but for the whole series."

The sixth episode was written largely by David Suzuki himself. Called *Improving on Nature*, it focuses upon a subject that is one of Suzuki's perennial favorites: eugenics. In one memorable scene Suzuki discusses what he sees as the origins of Second World War Nazi eugenics policies — the exaggerated scientific claims of a number of prominent American and European geneticists in the early part of this century. "It would be a terrible mistake," states host Suzuki, his face ashen as he stands in front of Dachau concentration camp, "to dismiss the Nazi death camps as an aberration of one madman and his fanatic followers. Race purification and improvement was state policy and widely supported. It had been expounded for decades, not just by third-rate intellects, but by some of the leaders in the scientific and medical communities."

Because it focuses almost exclusively on the darker side of the history of genetics, Suzuki predicted before it aired that the segment would be something "many of my colleagues will hate." The early days of genetics were "a very intoxicating period," he says. "But it's not just something out of the Dark Ages of science's past. I hear people saying *exactly* the same kinds of things today." The important thing for the public to understand is that "scientists are first and foremost human beings, subject to the same limitations of cultural bias as anyone else." In Suzuki's view, "demystifying science and showing that scientists are vulnerable human beings in the long run helps science rather than hurts it."

BORN IN VANCOUVER in 1936, Suzuki is a *sansei* — a third-generation Japanese Canadian. His memories of early childhood in British Columbia are mostly fond ones, filled with nature hikes and fishing trips with his father, an avid outdoorsman who inspired him to go into biology. But British Columbia's Asian population had, in fact, experienced years of racial tension, culminating in a series of race riots in Vancouver during the 1930s. And on February 26, 1942, in the hysteria following the bombing of Pearl Harbor, the Canadian government ordered the relocation of more than 20,000 residents of Japanese ancestry from the coastal area of British Columbia to internment camps in the province's isolated interior. Suzuki, his mother and sisters, and, later, his father were confined to an old mining town named Slocan for the duration of the war. Afterward, they were forbidden to return to their Vancouver home and reluctantly moved east to Ontario to begin a new life.

Suzuki sees the Slocan years as the most important period of his life. During that time white Canadians rejected him as Japanese and embittered Japanese-speaking internees rejected him as Canadian. By the time the Suzukis left Slocan, nine-year-old David despised his Asian heritage: "I wanted to have an eye operation, to dye my hair, and to change my name," he recalls. "I didn't even want to walk down the street with my sisters because I was embarrassed to be seen with other Japanese." The experience was, he says, "the most profound factor shaping my personality — filling me with self-hate, insecurity, and a drive to prove my worth to other Canadians."

Suzuki remembers beatings by classmates at Slocan, a sudden loss of interest in school, missing a father assigned to a "volunteer" Japanese road crew — little more. But one of his classmates in Slocan's Pine Crescent Elementary School remembers well. Then a young *nisei* girl named Joy Nakayama, Joy Kogawa is now a highly respected Toronto novelist and poet. Her award-winning novel *Obasan*, based on her own internment experiences, offers revealing glimpses into the fragmented childhood realm she shared with Suzuki.

Suzuki still finds *Obasan* too painful to read. Yet those years, says Kogawa, are crucial to understanding his deeply personal motivations to be a science celebrity in Canada. Suzuki himself says,

"People ask why I seem so driven to continually express such views. They say, Why you? Who do you think you are? Why do you think you should be doing this kind of popularization of science? And it's all just intimately related to the war years."

When Suzuki left Ontario to attend Amherst College in Massachusetts on a scholarship, he intended to become a physician. But during a genetics course in his junior year, he became utterly entranced by the elegance of gene-bearing chromosomes and eventually turned down an offer to attend medical school in Canada, enrolling instead in a doctoral program in genetics at the University of Chicago. Suzuki's studies there focused on subtle influences on chromosome crossover rates in *Drosophila*. But he was also becoming increasingly drawn into the surging American civil rights movement. He joined Asian-American political groups, participated in sit-in demonstrations with Chicago blacks, and later, as a researcher at Oak Ridge National Laboratory in the early 1960s, became the only non-black member of Oak Ridge's branch of the NAACP.

But the emotional price of this commitment proved high. Once, while driving with his young Japanese-Canadian wife through Alabama and Mississippi, Suzuki actually became physically sick with rage at the sight of a Whites Only sign. Gradually his visceral response to racism transformed his fury at Caucasians into "an obsession, a sickness." So, abandoning earlier ambitions of a prestigious research career in the United States, he returned home with his family to Canada. "When I left the States," he recalls, "I vowed I would never return, but I could see that science in Canada would be a much smaller puddle."

In 1962, as an assistant professor at the University of Alberta in Edmonton, Suzuki first tried his hand at teaching science on TV. It was a modest assignment — lecturing in a series of Sunday morning campus telecasts for about $50 a show — and it evoked little public response. He had more impact with a weekly CBC TV show called "Suzuki on Science," which he started hosting in 1971. Much of its content was typical science-TV fare — the inner workings of DNA's code, the mechanisms of immunity, evolution, and extinction. But the titles of a few programs — *Race and Intelligence*, *Chemical and Biological Warfare*, *Science in Vietnam*, *Women in Science* — also suggest the remarkable range of topics Suzuki already felt scientists should be willing to openly discuss with the public.

Eventually Suzuki's enthusiasm for laboratory research began to flag. His obsessive commitment to his lab, he felt, had contributed significantly to the breakup of his first marriage. And he was beginning to find growing satisfaction in his local media projects. He was also experiencing a nagging anxiety about his scientific work. Once in one of Suzuki's popular campus bull sessions with students — a passionate debate on gene cloning — a student challenged him personally on the ethics of his own genetic research. When Suzuki protested that he was merely experimenting on fruit flies, the student pointed out that even such seemingly harmless research went into a common pool of knowledge about heredity that could eventually be exploited to control or engineer people. The thought completely devastated Suzuki and served as a revelation. It represented, he recalls, "the most grotesque kind of juxtaposition between the two great loves of my life — genetics and civil rights. Suddenly they came together in the most horrific way." He began reading avidly about the history of eugenics, and for a year he came close to giving up his research entirely simply "because the idea of my contributing to knowledge that might be misused was just too horrifying."

By 1975 Suzuki had signed on as the original host of a CBC radio show called "Quirks & Quarks," whose weekly excursions into the lives and laboratories of scientists still remain a virtual Canadian institution. The year before, CBC had invited him to host a new TV series called "Science Magazine." And five years later, CBC merged "Science Magazine" with the perennial "The Nature of Things," establishing Suzuki as a resident scientist and mentor.

In 1977, Suzuki was invested as an officer in the Order of Canada, the country's highest civilian honor, for his science education efforts, joining the ranks of eminent Canadians such as the novelist Margaret Atwood and the jazz pianist Oscar Peterson. With his CBC colleagues, he's won several national and international prizes for science documentaries, including a United Nations Environment Program Medal for "Planet" and UNESCO's prestigious Kalinga Prize, which has been previously awarded to the likes of Bertrand Russell, Julian Huxley, and Margaret Mead.

Philip Morrison, an MIT physicist, a part-time TV scientist himself, suggests that no respected American scientist can come close to Suzuki in terms of total time logged talking science on television.

SUZUKI, OF COURSE, cannot claim all the credit for that massive quantity of air time. The large government-subsidized CBC network, a major player in the national TV market, simply has no counterpart in the U.S. Thus the high profile that science — and scientists like Suzuki — are accorded in Canada is in part the legacy of a long tradition of Canadian federal commitment to putting science high on the agenda of national television.

But even Suzuki's staunchest critics give him high marks for the part he has played in building on that tradition. Science journalist Stephen Strauss, who sharply criticized "Planet," nevertheless also believes that most people have tended to overlook the enormous impact Suzuki's success story has had in developing Canada's international reputation for excellence in communicating science to the public. "He blazed a trail for other people in the field. I think you could even make the argument," says Strauss, with a touch of irony, "that proportionally there are now more world-class science explainers in Canada than there are world-class scientists."

This is hardly the case in the United States, according to Fred Jerome, a media expert with the Scientists' Institute for Public Information in New York. Except for a few shining exceptions, such as certain outstanding episodes of PBS's "Nova" series, he says, American science TV tends to stick to a monotonous "gee-whiz" approach. It also is far too cautious in pursuing stories — in physics or mathematics, for instance — that while rich in ideas lack an obvious visual dimension. As for its willingness to tackle difficult ethical questions, Jerome says, "science programming in the U.S. has not even reached the point where that is an issue. There are simply too few science shows to judge. But what little there is has not been as probing as the print media's coverage of science."

Jerome feels that several factors are absolutely essential for a first-rate show. The first seems obvious: the need to breathe life into scientific ideas. He gives Suzuki high marks here, and even Suzuki's toughest critics praise his ability to effectively communicate not just scientific ideas but his own passion for science. James

Murray, the executive producer of "The Nature of Things" and a close personal friend of Suzuki's, captures this quality by saying, "People believe what he says, largely because he takes what he is saying so seriously — in a medium that often fails to take itself seriously."

Another of Jerome's requirements that Suzuki fills is the need for a familiar *human* element in the telling of TV science stories. Suzuki's shows openly embrace the human dimension of science, but they sometimes go even further. Unlike most TV scientists, Suzuki has repeatedly shown himself willing to bring details of his personal life to the screen. While filming a "Planet" segment dealing with the notion of death, for instance, Suzuki insisted on recounting his mother's recent death of Parkinson's disease, and his own excruciating pain at watching her life unnecessarily prolonged by artificial means. To him, it was quite natural to share such feelings with his viewers.

Science shows also ought to challenge viewers intellectually, says Jerome — stretching the limits of their stunted vision of science beyond mere memories of high school science classes. In Strauss's view, that is precisely where "Planet" excelled, whatever one's feelings about its ultimate message. "Most TV science programs *don't* have a lot of intellectual content," he says. "They *don't* make arguments. And when you reflect back on them you usually *don't* remember ideas or arguments. You remember images. I usually don't have the feeling after watching a TV show of wanting to engage in a discussion. After 'Planet' I did."

Quality science programming should also be willing to explore political and ethical aspects of science and technology that may prove controversial, Jerome maintains. Indeed, the gradual escalation of news coverage of science's daily impacts on viewers' lives compels science shows to do just that. Following "Planet"'s debut, few Canadians could deny Suzuki's bold efforts in this area. In fact, David Spurgeon, a veteran Canadian science writer and former *Globe and Mail* reporter, suggests that controversy is, intriguingly enough, absolutely vital to the Suzuki mystique. "David," he says, "is not afraid of controversy. In fact, he enjoys being a controversial figure. He cultivates it; he always has. And I think that has helped increase his credibility to the general public."

Although Suzuki may be giving science programming some of

the vitality and controversy it needs, his stance as science's self-proclaimed ethical standard-bearer doesn't always go down well with some colleagues. Once, years ago, while preparing a show on the nuclear arms race, as Suzuki tells it, he launched into a heated off-camera debate on the morality of scientists with a famous American physicist who, like many physicists of his generation, had worked on the atomic bomb project during the Second World War.

"Don't you feel," Suzuki queried, "that *you* and *I*, as scientists, have a special responsibility to society?"

"No," replied the physicist indignantly. "We're like any other citizen. And if I choose to speak out against nuclear arms, it is as an ordinary citizen."

"But don't you think you have a responsibility that goes *beyond* that of ordinary citizens on such matters?" pressed Suzuki. "After all, it's *our* colleagues and *our* students who are inventing weapons of war. They're *our* peer group, and we understand exactly what they're doing. We teach them, we communicate with them — surely that puts an *extra* ethical burden on us as scientists."

That proposition, recalls Suzuki today, absolutely infuriated the physicist, as it has many other scientists to whom he has posed it during his decades as a broadcaster. And perhaps it would infuriate our other "visible scientists" in Canada, the U.S., and beyond as well, many of whom seem to be growing increasingly reluctant to speak out vibrantly on ethical issues in science.

"I think that the willingness of scientists themselves to publicly raise ethical and political issues is decreasing," observes Goodell. "And that should be cause for concern."

1986

The Heart's Lonely Hunter: Canadian Cardiac Scientist Adolfo de Bold

ONE MIGHT THINK it would be easy enough to find Adolfo de Bold. After all, the 45-year-old Argentina-born pathologist's recent breakthroughs in heart physiology have earned him global acclaim, and his name is now frequently mentioned in the same breath as that of such monumental figures in Canadian science as Sir Frederick Banting — the Nobel laureate credited, along with three fellow Toronto researchers, with discovering insulin more than six decades ago. Indeed, de Bold himself is considered by many to be a leading contender for a Nobel prize.

And yet, on a crisp January morning at Ottawa Civic Hospital, where de Bold has been serving as the first director of a planned $17 million expansion of the prestigious University of Ottawa Heart Institute Research Centre since July 1986, no one seems to have heard of the man.

Eventually, a persistent visitor finds, in the shadow of the hospital, down a quiet, snow-covered lane from the noisy traffic of Carling Avenue, the old two-story red-brick building that serves as the temporary home of de Bold's new research facility. With its darkened windows, lumber-strewn stairway, and unmarked entrance, it at first appears to be abandoned. At best, it seems an unlikely setting for a scientist of de Bold's stature to launch what has been touted as a major interdisciplinary assault on heart disease. But the site, removed as it is from the mainstream, seems to typify a certain maverick quality in the career of a remarkable — and, in the minds of some members of the scientific community, quixotic — Canadian scientist.

A virtual unknown a decade ago, de Bold startled the scientific world in 1981 by detecting a previously unknown hormone. Astonishingly enough, the hormone was secreted by an organ that scientists had never even considered as an endocrine gland — the heart. De Bold called the substance the atrial natriuretic factor (ANF), an initially imposing label that refers to its origin in cells located in the upper chambers, or atria, of the heart and to its ability to stimulate the kidneys to excrete water and sodium, or natrium, rapidly from the body.

De Bold's discovery, along with his team's subsequent success in determining the precise chemical structure of the ANF molecule, catapulted him from relative obscurity as an assistant professor of pathology at Queen's University in Kingston, Ontario, to prominence as an internationally acknowledged pioneer in heart physiology. And because ANF has been found to play a central role in the body's natural mechanisms for regulating blood pressure, his finding has also ignited a global search on the part of major pharmaceutical companies for possible ANF-related drugs.

In the lobby of the dimly lit Ottawa building, an outdated tenant index on the wall lists neither the new heart-research center nor its eminent director. But upstairs, at the end of a long, echoing corridor — unadorned except for a simple bulletin-board display of photographs showing the dark, microscopic atrial granules that have been de Bold's lifelong quarry as a scientist — lights are ablaze and voices murmur. There, in a small office next to a well-equipped laboratory staffed by a handful of associates, including his wife and long-term research collaborator, Mercedes, de Bold slouches in a swivel chair. At his feet, ready to be hung, lies a gift from the hospital — a freshly framed *Maclean's* magazine cover bearing a miniature portrait of de Bold and 11 other Canadian honorees who "made a difference" during 1986.

He is a diminutive figure, and as he nods in greeting, he seems noticeably weary and to me somewhat older than his years. But as soon as I ask him to discuss the scientific importance of his recent discovery, de Bold quickly becomes animated and launches into the tale of his work. "The heart," he says, his words enriched by a Spanish accent, "has always been at the center of biomedical knowledge. You can trace it back to Aristotle, Hippocrates, Galen and William Harvey — all of them giants in science who had some connection with the heart. Their concepts led to the view of the heart as the organ in charge of the circulation of blood."

"What we now know is that the heart also synthesizes a hormone — ANF — that allows it to regulate the amount of work it has to do. The volume of blood reaching the heart controls the strength of the message that the heart sends — in the form of circulating ANF molecules — to the kidneys and other organs targeted by the hormone. They, in turn, act to reduce the volume of blood reaching the heart." De Bold adds enthusiastically: "Nature is terrific! You simply could not design a better system than that."

Before 1981, in what today might, tongue in cheek, be called the Dark Ages of heart endocrinology, or the science of cardiac hormones, scientists generally looked upon the human heart as little more than a glorified mechanical pump. An exquisitely designed one, to be sure, but one whose task seemed limited to a lifetime of rhythmic muscular contractions that would impel oxygen- and food-laden blood throughout the tubular labyrinth of the circulatory system. In 1981, that illusion was shattered. That year, de Bold, in conjunction with a team of physiologists at the University of Toronto, published a report on an experiment that would become a landmark in heart research. The researchers had injected ground-up heart tissue containing atrial granules directly into the bloodstream of a carefully monitored live rat. Stimulated by some potent ingredient in the heart extract, the rat responded dramatically. Its blood pressure plummeted, and its kidneys began to excrete prodigious amounts of urine laced with dissolved sodium ions. "It was," says de Bold triumphantly, "like turning on a water tap."

Within three years, de Bold and his colleagues at Queen's University crowned the discovery by announcing the precise chemical structure of the mysterious hormone stored in the heart's atrial granules: ANF is a peptide, a small proteinlike molecule consisting of 28 amino acids linked together like so many beads on a string. And it has been found not only in rats but also in organisms ranging from humans and other mammals to such lowly invertebrates as the clam.

Studies by researchers around the world have since demonstrated that ANF reduces blood pressure by counteracting some of the mechanisms that often go awry in victims suffering from various high-blood-pressure conditions. The hormone coaxes the kidneys to flush salt and water from the bloodstream. It reduces the secretion of renin — a hormone that helps to maintain blood-pressure levels — from the kidneys. It inhibits, in the adrenal glands, the release of aldosterone, a hormone that encourages the body to store sodium. And it also affects the muscular walls of blood vessels, the brain and several other organs involved in the body's intricately coordinated blood-pressure control system.

The medical promise of ANF arises, in part, from the fact that as a natural product of the human body, it appears to be able to reduce blood-pressure levels without causing many of the harmful side effects associated with the synthetic antihypertensive drugs currently on the market. In addition, an understanding of ANF's complex physiological role in the regulatory mechanisms in the body might also be exploited by pharmaceutical research for drugs to treat congestive heart failure, certain kidney ailments, cirrhosis, and other diseases.

Multinational drug companies in the United States, Europe and Japan are keenly aware of the potential rewards for practical ANF-related drugs and are eagerly searching for them. There is an estimated $5 billion global market for pharmaceutical products for high blood pressure and heart disease alone — a market of more than $100 million in Canada. Keenly aware that his work on ANF could lead to the development of lucrative ANF-related pharmaceutical products, de Bold has hoped from the beginning that the discovery might be used in ways that would economically benefit his adopted home — in his words, "to build something for Canada."

ADOLFO DE BOLD was born on Valentine's Day in 1942 in Paraná, a city several hundred miles north of Buenos Aires, up the Paraná River. The younger of two children, he was raised in a middle-class family. His father worked as a senior bureaucrat for the Argentine government; his mother, who had trained as a teacher, was a housewife. "For reasons that are beyond me," recalls de Bold, "I was interested in science from my earliest memories.

I vividly remember, for example, being very impressed when I was eight or nine years old by a picture of Louis Pasteur in my older sister's microbiology textbook." But the most important influence that his Latin upbringing may have had on his career as a scientist in Canada, he believes, was his parents' "slightly romantic view of life in Argentina, in which intellectual achievements were often seen as more valuable than economic ones."

In 1968, de Bold earned a degree in clinical biochemistry at Argentina's National University of Cordoba, where he also worked for a time as an undergraduate laboratory instructor with his future wife. His training, designed to license him to operate a private pathology laboratory in Argentina, was traditional, multidisciplinary and pragmatic — all qualities de Bold would bring to bear in his own distinctive style of heart research. The same year, de Bold and Mercedes traveled north to Canada, in his words, "with no plan whatsoever for what the future would bring." Both of them soon enrolled in graduate programs in the department of pathology at Queen's University. After a brief flirtation with insulin research, de Bold turned to a scientific problem that for years had piqued the curiosity of his graduate adviser, Sergio Bencosme, a pathologist from the Dominican Republic. De Bold began to study the clusters of microscopic granules found in the cells of the upper chambers of the heart.

As early as 1956, scientists had noted the presence of dense dark-staining granules in the atrial cardiocytes, or heart-muscle cells, of guinea pigs. Adrift in the fluid interior of cells, the minute globules bore a remarkable resemblance to the hormone-laden storage granules found in secretory cells of the pancreas, the pituitary and other endocrine glands of the body. But their role in the physiological affairs of the heart remained a mystery. It would be de Bold's challenge to try to find ways to isolate the enigmatic granules, assay their biochemical cargo and uncover their biological function.

A few years — and hundreds of thousands of homogenized rat hearts — later, de Bold had analyzed enough atrial granules to discover that they did not harbor, as many had suspected, a group of adrenaline-related hormones called catecholamines. De Bold was deeply disappointed. But, driven by his conviction that "nature simply doesn't put granules in a cell without giving them a very important function," he persisted with his project, despite gnawing doubts as to its worth. After finishing his doctoral thesis in 1973, he decided to accept a faculty research position at the nearby Hotel Dieu Hospital, which is affiliated with Queen's. It was hardly a glamorous position for a new graduate. The hospital had few researchers on staff and did not have a strong reputation as a research facility. But the position did have its advantages. It freed the young assistant professor from university teaching responsibilities so that he could continue to devote almost all his time to what was becoming a growing obsession with the atrial granules of the heart.

At Hotel Dieu Hospital, de Bold was assigned a small, windowless laboratory, located along one of the gloomy, tile-floored corridors that run through the basement of an old wing of the institution. He filled it with secondhand furniture and equipment scrounged from campus. "I had nothing — absolutely nothing," recalls de Bold. "I was *completely* isolated. There was no one to talk to, and hardly anyone at the hospital was doing research. But that is where my broad laboratory training from Argentina paid off. I didn't need anyone to help me with my research. It was a one-man show."

In retrospect, de Bold's solitary research experience at Hotel Dieu Hospital may have been a crucial ingredient in his later success. To some, it even stands as a classic case of the necessity of supporting seemingly unimportant basic research carried out by dedicated individualists on the fringe of more competitive, grant-oriented mainstream science. De Bold was not the only scientist who had displayed a keen interest in atrial granules. Over the years, researchers at Yale University, the University of Paris, the Clinical Research Institute of Montreal and other institutions around the world had scrutinized the particles for clues, only to abandon the search in frustration after a time. "Many picked it up for a while, then dropped it," says de Bold. "It wasn't very productive research, so they didn't stick to their guns. I don't think anyone managed to stay with it continuously as long as I did."

Working essentially alone, despite occasional collaborations, de Bold was free to try tedious, often unrewarding, technical approaches that might well have been judged unproductive, even eccentric, in a more conventional research setting. "Solitude," muses de Bold, "does offer a certain freedom. If I had been working in any other environment, I really don't know if I would have discovered ANF. I probably would have abandoned atrial granules to devote more time to other problems." With a mischievous grin, he adds, "Of course, I am stubborn, too."

During the next few years, de Bold and others added a few new pieces to the puzzle of atrial granules. Assisted by a series of modest research grants from the Heart and Stroke Foundation of Ontario, the Medical Research Council and other agencies, he developed an improved technique for counting the unevenly dispersed clouds of particles in each cell using a light microscope. Inspired by a suggestion from Mercedes — who by 1975, following the birth of the first of their five children, had begun to collaborate with him in the laboratory — de Bold concocted a lead-based dye that allowed selective staining of the granules. Then, to count them, he superimposed a rectangular grid over the field of dyed granules contained in a sample of heart tissue. In so doing, he was able to monitor accurately the shifting fortunes of atrial granules in heart cells of laboratory animals exposed to a wide range of experimental insults. "Until then," says de Bold, "it had been very difficult to count the granules. You see, the cells in the atria are not an orderly bunch, like those in the ventricles."

Armed with his improved technique, de Bold set out to test a number of hypotheses advanced over the years by several scientists concerning the possible function of atrial granules. Among them was the notion, based on less reliable granule-counting methods, that the structures might be somehow involved in the regulation of water and sodium levels in the body. In 1978, de Bold experimentally confirmed a French study that had reported that the number of granules in the atrial heart cells of laboratory rats increased when their diets were deprived of sodium. The finding suggested that the granules might be acting as cellular storage depots for substances involved in sodium balance. If that were true, reasoned de Bold, the logical target would be the kidneys — the elongated, paired blood-filtering organs that are nestled against the spine and that actively regulate water and electrolyte levels in the bloodstream.

In 1979, after attempting unsuccessfully to demonstrate a connection between atrial granules and renal function, de Bold scanned a Heart Foundation annual report for the name of a local expert in kidney function. He chose a University of Toronto renal physiologist named Harald Sonnenberg — a fortunate choice. Sonnenberg had already spent years searching for an elusive, hypothetical "natriuretic hormone" that many scientists believed must be involved in blood-volume regulation. The "missing link" in the endocrine system had been postulated as early as 1935 to account for otherwise inexplicable fluctuations in the blood's sodium and water levels. But no one had been able to confirm its existence. John Ledsome, chairman of the Department of Physiology at the University of British Columbia and a leading ANF researcher, explains: "In many ways, the heart seemed like the last place to look for it. Most of us figured that the heart already had enough to do."

Anxious to test his atrial-granule hypotheses, de Bold telephoned Sonnenberg and invited him to take a "shot in the dark" with him by conducting the crucial experiment. De Bold, it was agreed, would supply a sample of pulverized heart tissue containing an abundance of atrial granules. Sonnenberg, for his part, would arrange to inject the extract directly into the bloodstream of a normal rat and record its effect on the kidneys. Their first experiment was a dismal failure. De Bold's sample, inadvertently contaminated with enough potassium to cause cardiac arrest, killed the rat. But a second trial a few months later revealed ANF's now legendary capacity to flush sodium and water from the kidneys and at the same time lower blood pressure.

Despite the dramatic response, de Bold had considerable difficulty finding a reputable science journal willing to publish a full report of the landmark experiment. The problem was twofold. First, like most major departures from the scientific status quo, de Bold's discovery of a hormone in the heart faced a measure of conservative resistance in some parts of the scientific community. Second, it suffered from an unfortunate choice of words. Because de Bold referred to the substance as a natriuretic factor, some of his peers viewed his study as just another search for the missing natriuretic hormone — tantamount to hunting for a hormonal unicorn. The full report finally appeared in a 1981 issue of *Life Sciences*, but only after being roundly rejected by another publication. "The paper was sent to *Life Sciences*," says de Bold, "because it was rejected by a more prestigious journal in the United States.

They simply did not believe it. Because we referred to ANF as natriuretic, it was apparently equated with the mythical natriuretic hormone and fell into an area of science that was not considered very rewarding. It was a problem of semantics."

When they had read the report and confirmed its findings, however, most scientists in the field realized that de Bold was on to something big. De Bold himself remained guardedly skeptical of the initial results. He was greatly encouraged by them but simply considered it too soon to celebrate. "For me," he says, "there is always an alternative explanation. There is no single moment when you can say Eureka! You must always try to shoot down your hypothesis. In this case, the crucial thing was to isolate ANF, determine its structure, synthesize it, then inject it back into the rat and stimulate the identical response. At that point, I would be convinced we had something."

De Bold knew that if he was to have any hope of crowning his initial discovery of ANF by determining the precise chemical structure, or sequence, of the molecule, he would have to move quickly to assemble an interdisciplinary team of experts at Queen's. But ironically, just when he most needed financial support, it was not there. "The funny thing is," he says, "the year we reported the discovery of ANF, the granting agency must have missed the publication. They gave me a terminal grant."

SNOWFLAKES FROM a January storm swirl outside Geoffrey Flynn's office window as the Queen's University protein chemist recounts his brief, but successful, collaboration with de Bold on sequencing the ANF molecule. Eager to maintain the momentum of his initial discovery, de Bold had approached Flynn in 1981 at the same time that he contacted the campus patent committee. Clearly, any future claims by the university for patents relating to ANF would only be strengthened if Queen's researchers could become the first in the world to publish the correct sequence of the hormone. "De Bold needed a protein chemist. He had to have one," says Flynn excitedly, eyes bright behind dark-framed spectacles. "And I thought it was the most wonderful opportunity that I had ever heard of as soon as he told me about it."

From the start, however, the project was crippled by a shortage of funds. De Bold was already gaining recognition in some circles as the discoverer of ANF. Yet he seemed unable to win sufficient institutional support to proceed with his ground-breaking studies. Because of his decidedly independent approach to science — along with an open disdain for research aimed "more at satisfying the needs of grants than at satisfying the difficult intellectual questions posed by a problem" — he had only limited resources of his own. In the eyes of some fellow scientists, de Bold simply lacked the track record as a researcher that would warrant a major show of support. Nor, by his own admission, did he possess powerful allies in the scientific establishment who might come to his aid.

De Bold views the obstacle course he faced to obtain financial support as less a personal problem than a Canadian one. It reflects, he believes, the failure of Canadian institutions to recognize quickly the value of homegrown scientific ideas and to provide the funding and counsel they require to assure a Canadian role in their future commercialization. "In Canada," says de Bold, "we simply do not have a system that can respond quickly enough to a breakthrough like the discovery of ANF. I went everywhere. But perhaps I wasn't smart enough. In retrospect, I should have gone to the highest authorities in the land. There was just no one willing to stick out his or her neck to give us the money when we needed it."

He pleaded with Queen's faculty and administrators for help in finding short-term financial assistance so his team could continue to compete in what was fast becoming a hotly competitive international race to unmask the ANF molecule. They responded by offering de Bold and Flynn $36,000 — in the form of a loan from a campus principal's fund normally used to support faculty members early in their careers. The researchers also managed to scrounge a modest two-year grant of $29,000 from the Medical Research Council, designed to allow de Bold to extract and purify ANF and Flynn to sequence it. "This was *after* Adolfo had done the pivotal experiment on ANF," groans Flynn. "It was pitiful."

To make matters worse, Flynn's high-performance liquid chromatograph, a laboratory device crucial to the analysis of ANF, suffered a series of breakdowns. So the scientists resorted to moonlighting. They worked feverishly at the Hotel Dieu Hospital to run their tests on a similar machine that was available only evenings and weekends. Yet despite their limited resources, compared with ANF-research teams in Montreal and abroad, the Queen's scientists had at least one advantage. De Bold — who prides himself above all else on a perfectionist's attention to detail in laboratory technique — had pulled ahead of the pack by developing, with the help of Mercedes, an innovative method of using acetic acid to leach ANF gently from the atrial granules. "And that," says Flynn, "was quite a clever move by the de Bolds."

Meanwhile, alerted by de Bold and Sonnenberg's report, several teams of scientists in Canada, the United States and Japan had already committed substantial resources to winning the ANF-sequencing contest. The leading Canadian contender, headed by Marc Cantin, was based at the Clinical Research Institute of Montreal. Scientists at the institute had actively engaged in research on hypertension-related problems for more than three decades. According to Jacques Genest, scientific director of the institute at the time, the Quebec researchers succeeded in confirming de Bold's initial experiment within 24 hours of receiving word of it in 1980. Cantin himself had studied atrial granules intensively during the 1970s, even collaborating briefly with de Bold and Bencosme. In his laboratory at the Clinical Research Institute — today a beehive of ANF-related research activity, in contrast with de Bold's more sedate laboratory in the Heart Institute Research Centre — Cantin vividly recalls reading the brief 50-line abstract of de Bold and Sonnenberg's experiment in 1980. "This is it," he thought. "This is it."

The Queen's University scientists seemed hopelessly outclassed by the Montreal team in their race to sequence the ANF molecule. Not only was Cantin's group much larger and well funded, but it had also managed to negotiate solid corporate backing in the form of an agreement with Merck & Co., Inc., the huge multinational drug firm based in Rahway, New Jersey. Merck was eager to provide financial and technical support for ANF research that might lead to the development of successors to its highly lucrative antihypertensive drug marketed as Vasotec. The pact with Merck, one of the most glamorous and profitable of the big multinationals — in 1986, it had revenues of $4 billion (U.S.) and a research and development budget of $460 million — provided the Clinical Research Institute with a level of financial and technical support for its ANF studies that was simply unavailable in the small Canadian pharmaceutical industry. An American research team in St. Louis, Missouri, adopted a similar funding strategy, forming an ANF alliance with the giant chemical conglomerate Monsanto. And others followed suit — but not de Bold.

By early 1983, Flynn had mapped 24 of the 28 amino acids now known to make up the ANF peptide. But to complete the task quickly, he needed expensive, state-of-the-art laboratory equipment and funds to bring in Queen's scientists with expertise in molecular biology and immunology. Armed with their preliminary evidence, he and de Bold again solicited support from campus officials, arguing that Queen's stood to benefit handsomely from licensing royalties if their sequencing effort succeeded. "By this time," recalls Flynn, "Queen's University officials were buying ANF as a genuine scientific breakthrough. But it had taken them a long time. They had shown incredibly little response earlier."

Together, the researchers also applied for financial support from IDEA Corporation, an Ontario Crown corporation designed specifically to facilitate the rapid commercialization of new technologies. Before long, they received $800 000 for a two-year period plus $400,000 from the Natural Sciences and Engineering Research Council. By late 1983, bolstered by the infusion of funds, Flynn and de Bold had overcome a series of discouraging setbacks to identify the remaining four amino acids in the ANF formula. They had also managed to detect a critical feature on the landscape of the ANF molecule — a bridge of sulphur atoms connecting two segments of the peptide chain. It would prove to be not only vital to the physiological activity of ANF but vexing to some of their closest sequencing competitors.

Finally, on December 28, 1983, Flynn and Adolfo and Mercedes de Bold published their formula for rat ANF. Just days later, on January 6, 1984, a team of American scientists published its version. And later that spring, Cantin's group at the Clinical Research Institute of Montreal published its findings. Oddly enough, not all of the ANF sequences contained the same number of amino acids. Only later did it become clear why. When synthesized in the atrial granules of the heart, the hormone is 152 amino acids long. Soon after, however, special enzymes whittle it down to a lean circulating version that is just 28 amino acids long. The de Bolds and Flynn were the first to report the sequence of the active blood-borne hormone. The molecule's structural idiosyncrasies, including its distorting disulphide bridge, also had misled some scientists into sequencing fragmented forms of ANF.

DEVELOPMENTS IN ANF research continued at an explosive pace through 1984. Early that year, for example, researchers successfully sequenced the human ANF molecule — a 28-amino-acid peptide that differs from rat ANF by only a single amino acid. Later, recombinant-DNA techniques were used to clone both rat and human ANF genes inside yeast and bacteria cells. The creation of a cheap, abundant supply of synthetic hormone opened up ANF research to hundreds of scientists worldwide. But de Bold, whose painstaking studies of ANF now seem to be increasingly dwarfed by the hundreds of reports arising each year from major institutions, still relishes the sequencing victory in 1983 as another tiny vindication of "small science" over "big." Says de Bold: "Research should be fun. It is a contest between you and nature — to see who outwits whom. Against all odds, we won the race to sequence ANF. We beat everybody by spending those nights at Hotel Dieu Hospital."

But even with the precious formula for the cardiac hormone he had discovered in hand, de Bold faced frustrations in trying to realize his dream of securing a prominent Canadian role in any future commercialization of ANF. With the help of Heino Lilles, a Queen's University law professor and administrator, as well as other campus officials, de Bold continued to approach pharmaceutical companies in the hope of negotiating a licensing agreement that would both assure the continuation of his own studies on ANF and win long-term financial benefits for Queen's. He contacted small Canadian firms — including, ironically, Connaught, the drug company spawned by the limited Canadian commercialization that resulted from Banting's work on insulin at the University of Toronto. But inevitably, Canadian research and development programs either were too scientifically specialized to adapt to ANF drug studies or were simply nonexistent.

So most of de Bold's efforts were directed, by necessity, toward mammoth foreign-based corporations that had the resources to lure top scientific talent, build lavishly equipped research facilities and invest the estimated $100 million (U.S.) required to develop, test, and manufacture a new drug. They included names like Ciba-Geigy, Upjohn and Du Pont. "Some people might have thought that we couldn't exploit the discovery here in Canada because of our lack of major pharmaceutical companies," says de Bold. "But of course we could. We could sell it or at least collect royalties on it. It was my hope that in this way, the ANF discovery would become a Canadian centerpiece — one that would focus greater attention on the urgent need for us to improve our ability to benefit from our own scientific discoveries."

Over the next few years, the university's courtship of pharmaceutical companies suffered a long series of rejections. In some ways, the pattern seems predictable. Unlike their counterparts at such elite American technical institutions as Stanford University and the Massachusetts Institute of Technology, Queen's negotiators had little experience in bargaining for high stakes with big multinational drug corporations. They were also seriously hampered by a lack of adequate staff and funds to permit them to market their proposal systematically throughout the international pharmaceutical industry. Limited financial support and counsel from IDEA Corporation, in exchange for a provincial share in future ANF royalties, helped. But it was not enough. "I was Queen's point man on ANF," declares Lilles. "I think I spent more time with Adolfo over those three years than I did with my own wife. But our resources were very limited. What we needed was an effective technology-transfer organization at the university that had the necessary resources. We tried everything we could with the resources we had. But they were limited, and time was short."

In one fairly typical incident, Lilles and de Bold rushed to Toronto to meet with executives from a major Swiss multinational drug company about a promising ANF alliance. The negotiations seemed to be progressing well until their two corporate counterparts suddenly exchanged glances, then abruptly excused themselves from the room. After conferring privately for several minutes, they returned to announce that they had changed their minds: they felt they had sufficient in-house resources to pursue ANF research without de Bold or Queen's University. "It was very discouraging," says Lilles, sighing. "The firm's research director decided that they would keep the work in the United States. They would not bring it to Canada. And they would thereby avoid making any long-term commitment to anyone in terms of royalties."

Finally, in late 1986, Queen's negotiators did seal an ANF deal, although the details remain largely confidential pending a future joint public announcement. But in the process, de Bold, who holds patent application on ANF, and Queen's University, which possesses commercialization rights, had bitter tactical disagreements, the effects of which still linger. "I lost sight of what Queen's was doing with the ANF patents," says de Bold, "and I didn't think that was fair." Lilles, in response, insists that de Bold recently has been kept in the dark about the marketing of his scientific discovery only because, as research director at the Heart Institute Research Centre, he no longer holds a regular faculty position at Queen's. "I have a mandate to generate more research at the university as a part of this licensing activity — but not for the Heart Institute directly," says Lilles. "So I don't feel compelled to rush out to tell Adolfo everything we are doing."

Queen's University's new corporate partner, says Lilles, is a company based in Japan, a country whose $25 billion (U.S.) pharmaceutical market is second only to that of the United States. "Queen's, through its Canadian Enterprise Innovations Centre, has successfully licensed ANF," says Lilles proudly. "The agreement was signed December 10, 1986." In exchange for licensing rights to patents arising from de Bold's purification and sequencing of the rat ANF peptide, along with other unstated concessions, the Japanese firm is more than willing to provide long-term support for Queen's researchers. "The first thing they said to us," recalls Lilles, shaking his head in frustration, "was Why didn't you guys approach us three years ago?"

Despite the astonishing progress in the clarification of the nature of ANF since the beginning of the decade, the future of the cardiac hormone remains clouded. Complex questions surrounding the legal status of competing patent applications on ANF-related molecules are still hotly controversial and unresolved. Queen's University claims a patent application dated February 1983 for de Bold's 28-amino-acid rat peptide. The Clinical Research Institute of Montreal maintains that it has patent applications, filed jointly with Merck, dating from as early as August

1983 — though for molecules subtly different from de Bold's. Most of these and other applications await formal patent review and possible litigation. There is a consensus that depending on whether future ANF-related pharmaceutical products are derived directly from the ANF molecule or not, valid patents might be either extremely valuable or virtually worthless.

While the legal issues are being thrashed out, research on ANF is rapidly proceeding on many fronts. Serious obstacles remain, for example, before de Bold's discovery of ANF can be transformed into any major medical treatment. For one thing, ANF is a relatively short-lived hormone in the bloodstream, so a victim of hypertension might require in excess of a dozen intravenous injections each day. And, like insulin, it is a fragile molecule — easily rendered inactive by digestive enzymes if taken orally. So a number of pharmaceutical firms are now working on schemes to build a molecular armour around the peptide or, alternatively, to build ANF-like imposters, called analogues, that can serve as more effective drugs than the original. In the end, there seems to be a consensus among ANF researchers that years of study will be required before a broadly useful ANF-related pharmaceutical product is likely to reach the market.

For his part in unraveling the mysteries of the ANF hormone, de Bold had already won honors ranging from Alberta's Ernest C. Manning Principal Award and a Gairdner Foundation International Award, shared with Flynn and Sonnenberg, to the first Achievement Award offered by the American Society of Hypertension. And it is not uncommon to hear suggestions, from both his friends and his foes, that de Bold is a likely Canadian candidate for a future Nobel prize because of his pioneering achievement.

A few heart researchers, convinced that de Bold's discovery may be the key to unlocking a number of perplexing cardiovascular diseases, have, perhaps prematurely, publicly hailed ANF as the "insulin of hypertension." In fact, despite obvious differences in social and scientific settings, there are a few illuminating parallels between Banting's and de Bold's hormone discoveries. For one thing, both arose from what many consider to be unlikely quarters of the Canadian scientific community. In this regard, they stand as compelling evidence in support of an observation by J. J. Brown, author of *Ideas in Exile* — a classic study of the history of Canadian invention — "that the great breakthroughs have come not from the famous institutional research laboratories but from obscure and relatively unsubsidized individuals."

Banting's and de Bold's experiences also reveal some of the obstacles faced by those who attempt to commercialize major medical breakthroughs in a country with limited resources for, or experience in, successfully playing the aggressive global pharmaceutical game. "By and large, innovation in Canada is not a happy story," writes Brown. "Canadians have a good record for producing brand-new ideas, but as intelligent consumers of new ideas, we leave much to be desired. Time after time, we have gotten there first after a magnificent sprint and then stood around idly for years waiting to collect the risk capital required to get an industry going. Usually by the time we have solved the financial problem, other less torpid nations have caught up with and passed us."

Adolfo de Bold would certainly agree. "You need people who really *understand* what a discovery of the magnitude of ANF entails," he says. "It is only natural. A scientist cannot be a prophet of his own discovery. Yet in a way, adversity often helps — I have no doubt about it. In any case, I am happy that I did what I did. It was the work of a professional scientist — nothing more."

1987

Part III

ON NATIVE CULTURES AND NATURE

Introduction

What is man without the beasts? If all the beasts were gone, men would die from great loneliness of spirit, for whatever happens to the beasts happens to man. All things are connected. Whatever befalls the earth befalls the sons of the earth.

Letter from Chief Seattle,
Dwamish Indian leader,
to U.S. president Franklin Pierce, 1855

AFTER MORE THAN seven years in university settings studying the biological structures and functions of nature and many more than that reading and writing about them since, I have yet to encounter a more crystalline phrase capturing the very essence of what I have learned than Chief Seattle's "All things are connected" (however much his original words may have been embroidered by others over the years).

As many observers have already noted, this ringing statement is capable of bearing far more weight than imagined by those who prefer to dismiss it as the last romanticized ramblings of some nineteenth-century American "noble savage" long forgotten. As any thinking biologist knows — or should know — Chief Seattle's succinct four-word sentence resonates with all sorts of rich, documentable biological truths.

Consider a few brief examples. Evolutionary biologists confirmed long ago that *all* forms of life on our planet are in fact genetically *interconnected* — and are thus, in the truest sense, biological kin — by the common thread of a shared evolutionary past. Each of us carries evidence of this ancient, unbroken bond of kinship in specific DNA sequences or mechanisms of metabolic machinery we possess in common with other living species. Similarly one need only thumb through any introductory ecology textbook to begin to grasp that the activities of living organisms within, say, an Amazonian tropical rain forest or a Pacific coastal ocean ecosystem are also ecologically *interconnected* in countless ways — in the one-way flow of energy from sun to food-chain terminus, for instance, or in the endless cycling of water or minerals or carbon atoms.

There's more — much more, I think — common ground between the scientist's abstract, rationally coherent visions of the biosphere and the parallel deeply personal and unabashedly sacred visions of that *same* world offered by native elders. But that's another story. Suffice it to say that I nurture a growing sense that the West has much to learn from the accumulated ecological wisdom of First Peoples whose lives and cultural traditions have emerged from a daily intimacy with the natural world — the personal intensity of which most of us can scarcely hope to imagine. This is not to suggest that as a result we ought to suddenly abandon hard-won scientific insights or technological wizardries for a neo-Pleistocene existence or uncritically embrace *every* element of traditional aboriginal ecological perspectives — two common, knee-jerk reactions to this view. There is, I think, genuine substance to the notion that it's not too late, despite the tragic history of our dealings with indigenous peoples, to begin to pay attention to those aspects of native nature wisdom that *complement* and enrich the Western scientific worldview.

A little book titled *The Wintun Indians of California*, which I wrote as a graduate student in the early 1970s, is a record of my first awkward attempts as a fledgling biologist to try to explore native American cultures firsthand. I wrote it while working as a naturalist in the Shasta-Trinity National Forest, near Mt. Shasta, in Northern California, where I had the good fortune to befriend a number of elderly Wintu, members of the northern Wintu portion of the great Wintun nation. The selection called "Ways of the Wintu" is excerpted from that book, which, in turn, depended heavily upon ethnographic studies of the Wintu people written for the most part in the first decades of this century. It was intended to serve simply as an evocative "sketch," as I said in its preface, "of a Wintu microcosm, bound with the passage of a single solar year."

The *Natural History* article entitled "Flora, Shaman of the Wintu" arose from that same experience. It is a biographical account of one of the most remarkable women I have ever known — a highly respected Wintu shaman, or ecstatic healer (with no apparent successor), who gradually became a dear friend of mine. In a very real sense, she is its co-author. As you will discover, this is no "tale of power" *à la* Carlos Castenada. Rather it represented a cautious, scientifically self-conscious attempt on my part to try to reconcile the "separate realities" of the biologicial science of my academic training and certain traditional native American values to the extent that I then understood them. I have yet to fully complete that reconciliation in my own mind.

The article "Weavers of Wood," another *Natural History* piece, is a historical, ethnographic account of the exquisitely refined traditional Haida art of basketry. I researched and wrote it while living and teaching for a time in rain-swept Ketchikan, a stark and isolated little island town of surpassing natural beauty located in the Alaska Panhandle. Sadly, it also seemed to have one of the highest densities of bars and confirmed barflies in rural North America. I eventually fled Ketchikan, but not before interviewing, as part of a U.S. National Endowment of the Arts video project, several respected Haida basket weavers living both there and on British Columbia's Queen Charlotte Islands.

The article still holds a special importance to me for two principal reasons. First, it concerns a people — the Haida of northwestern British Columbia — who, in their continuing struggle to gain legal recognition of their traditional domain of Haida Gwaii, the lands and seas of their ancestral Queen Charlotte archipelago, have emerged as one of Canada's most articulate, farsighted and courageous native communities. And second, it attempts to grant Haida women artists a measure of the same sort of respect from the outside world as Haida men, as carvers and painters, received long ago.

Finally, the short essay "*Talaiden:* A Sacred Law against Laughing at Animals" offers a glimpse into my ongoing explorations of Native ecological perspectives embedded in traditional memory and myth. This one stems from the long-suffering Chewong people of peninsular Malaysia who — their lands threatened — are now living in extreme poverty and peril. To me, willingness to take such aboriginal notions of the sanctity of nature very seriously — and to *learn* from them — is hardly, as some critics would claim, a regressive retreat to "primitive" mysticism. Rather, it offers a potential wellspring of wisdom, often compatible with science, about the proper place of *Homo sapiens* in nature. And it underscores once again that the worldviews, human rights, and legitimate land claims of aboriginal peoples simply must be accorded the respect that has been so long and cruelly overdue.

Ways of the Wintu

The Wintun

A map of early California Indian tribes shows the Wintun-speaking peoples occupying a vast territory including most of the Sacramento Valley west of the Sacramento River, and extending from the southerly drainage of Mount Shasta to the northern shores of San Francisco Bay. Subdivided into three principal groups of related dialects — the Wintu (Northern Wintun), the Central Wintun (Nomlaki, or Wintun proper), and the Patwin (Southern Wintun) — the Wintun tribe as a whole was the largest in population and territory in northern California.

To realize the difficulties that anthropologists face in defining meaningful divisions of California Indian groups, the reader must recognize the unusual diversity of language and custom that existed in this region before the first influences from the Old World were felt. He or she needs to imagine the following:

(1) The landscape of an aboriginal California, rich in range of habitats and abundant in food resources

(2) A native population estimated at about 300,000

(3) An as yet uncounted number of language groups, or tribes, representing some seven distinct language families

(4) A socio-political organization that divided most of the native population into at least 500 discrete tribelets, each consisting of an association of one or more village settlements with its own surrounding food-gathering territory, and usually lacking any peculiar dialect or identifying title beyond one of its village names or a geographical description.

With consideration of these and other factors, we begin to perceive the fragile and elusive scheme that emerged in California over the estimated 11,000 to 40,000 years since, as some anthropologists claim, Stone Age man first followed caribou, musk ox, and mammoth across a temporary Bering land-bridge into North America. The route, exposed by the siphoning of ocean waters into accumulating continental glaciers, may have provided footing for several waves of Asian Paleo-Indians, further compacting California's cultural and physical diversity.

Despite obstacles that tend to blur our images of California Indian peoples, on a very large scale there are entire geographical areas that tend to adhere to a basic cultural heritage — the so-called ethnic provinces of northwestern, central, and southern California.

The northwestern Indians had rituals of wealth display, and a social system seldom extending beyond kinship bonds and relying heavily on rule of law. Central California was dominated by tribelets with distinctive religious cults and cosmologies. In the southern portion of the state were groups possessing a limited agriculture and sometimes nationalistic tribal organization. It was to the sprawling central California cultural complex that the Wintun and most of their neighbors were allied, and the Patwin (or Southern Wintun) are accorded an important role in the development and dissemination of some of its most prominent achievements.

Through the thousands of years of their existence the Wintu discovered and mastered the life-giving forces locked within their small world. Every feature of the landscape came to hold spiritual meaning and to be intimately known by each of them. Each animal and plant was endowed with name, aura, and practical usefulness; each was addressed with humility in deference to its place in the unbounded reality of which man is only one small part.

Within the crucible of their rugged and sun-drenched land the Wintu people forged a delicate symbiosis with their surroundings that would long stand the test of time.

***Oltipa* (Spring)**

The moon — silvery underside of the sun's belly — is fading from sight. An amber dawn spills over the peak of Chas-chim Puyuk

(Minnesota Mountain) and across the sky. The air is crisp and still, smelling of pine needles and moist earth; there is the steady rush of Winnimem's chill waters. Through the canopy of digger pines and oaks silently rises a thin wisp of smoke fed by a single campfire in the ancient village called Dekkas, sleeping below.

With the first light a figure appears, crouching, beside the fire, and a child's cry gently tears the stillness. Low voices follow, rising, as forms emerge from the cluster of bark huts.

Before the sun clears the mountain, villagers can be seen kneeling beside the river as they offer prayers to Sas, the sun, or to Olelbis, Creator of the Wintu people, appealing for good health, successful hunting and happiness in the coming day. Then, splashing or wading into Winnimem or its tributary, Dekkas Creek, which flows past the village, they bathe and swim in the slow currents.

It is *oltipa*, springtime, in the land of the Wintu, as it has been countless times before in the great cycle of seasons that imposes a rhythm upon these people as it does upon all living things. It is time to escape the cramped and smoky spaces of the family bark house (*krewel*), to gather clover, harpoon salmon of the spring runs, and feel the yellow warmth of the sun that the chill winter denied.

Some 50 Wintu, most of them related, inhabit the village settlement with its sprawling expanse of hillsides and canyons where foods and materials are gathered. Together, this semi-permanent living site and its surrounding resource-territory along a portion of the McCloud River drainage accommodate the Dekkas tribelet — a small but self-sufficient political organization, surrounded by scores of other similar ones. Besides the dozen conical bark dwellings — pole frames against which are laid slabs of incense cedar bark and evergreen boughs — two additional structures are visible at the downstream end of the village. One is a crude, brush shelter, the *yeltum krewel*; it is a simple place for women of the village to isolate themselves during menstruation. More impressive, a dome-shaped earth lodge or *altershlut* rises beside the river's edge. Owned by the village chief, it is almost 20 feet in diameter. Constructed by erecting a raised framework of log rafters radiating from a single center-post over a three-foot-deep circular pit, it is entered through the large smoke-hole in the roof. Once covered with an insulating layer of brush and earth, it provides comfortable quarters for unmarried men of the village during the colder months. It serves as village sweat house for men and, on occasion, houses ceremonial gatherings such as the annual initiation dance for Wintu shamans.

The village sits on a grassy flat above the junction of Dekkas Creek and the McCloud River, fringed by virgin stands of western yellow pine. Buckeye, redbud, manzanita, occasional digger pine, and other vegetation flourish, but because the villagers burn dead wood in their campfires and regularly control the growth of brush with ground fires, the surrounding forest is open and easy to travel through.

Here, as in most human societies, the physical simplicity of the village can be deceiving. It rests upon an invisible web of relationships, spun of kinship and economic bonds, which are sanctioned by the supernatural forces of the Wintu world. The village polity, with its interwoven controls and finite territory, constitutes the fundamental political, economic, and social organizational unit of Wintu society within which most of a Dekkas man's or woman's life will pass.

The tribelet is founded upon an ancient, widely accepted, and orally transmitted code of conduct and justice. The elderly are entitled to respect and care. Each man must develop skill as a hunter and fisherman to provide for his family; a woman learns to weave baskets and to prepare the food that she and her husband gather. Crimes such as murder or intervillage feuds demand either just retribution or monetary compensation, thereby neutralizing simultaneous urges for revenge and punishment.

Each village has as its recognized overseer and enforcer of tribal law, the village headman or *wi*. His position is usually hereditary, passing from father to son, but a *wi* who lacks his people's respect can be replaced by a more popular, and therefore more effective, man. Skill in oratory and arbitration are essential, for his daily duties will range from the sacred to the most mundane. Above all, a *wi* must possess a firm, democratic bearing — the language of the Wintu makes clear that their chief does not "rule" the people of a village, but rather "stands by" them.

The headman sets the opening day of fishing season and organizes communal deer hunts. He apportions acorn-picking territories among the villagers, and is often asked to distribute shares of food within this cooperative economy. He also is responsible for calling together members of his own and neighboring villages for feasts and ceremonial occasions, during which he delivers lengthy, dignified addresses that are laced with wise advice.

As a privilege of office, the *wi* is freed from much of the routine labor of gathering and preparing food. He often is granted a status that may carry with it considerable wealth and several wives. But it is not difficult to imagine that his responsibility of leadership brings personal burdens. A *wi* might agonize for long hours in the village sweat house over a decision affecting those closest to him.

Taboos, etched at an early age into each child's mind, also help to control and insulate activities within the village society. These restrictions are neither quaint nor arbitrary to the Wintu, but express an unspoken conviction that a person's every act or experience, however minute within a boundless cosmos, reverberates throughout his universe and therefore entails a profound sense of responsibility.

A woman, once initiated into adulthood, must eat and sleep in isolation from menfolk and abstain from meat, salt, and grease during her menstrual period. A young Wintu boy must bathe after shooting his first deer or catching his first salmon, and must not partake of this first flesh if he is to become a successful hunter and fisherman. Girls must undergo a ritual fast upon reaching puberty, and are taught to be shy, to attract their mates through such traditional marks of beauty as long hair, pierced ears, or chin tattoos of three vertical lines. A man must not touch or joke with his mother-in-law, or even stand close to her.

A disregarded taboo usually provides its own punishment. The irreverent man who comes in contact with his mother-in-law will be torn to shreds by a grizzly bear — a creature thought to harbor ghosts or evil spirits. Any woman who ignores the taboo against fishing during her monthly confinement may one day be possessed by the spirit of the suckerfish (Catostomus) and suffer a permanent contortion of her face. A thief will perhaps encounter a silent hostility from his neighbors, compelling him to leave the settlement forever.

We have glimpsed a portion of the social circuitry within the Dekkas village, but it is more difficult to see its occasional extensions beyond the tribelet boundaries. For as far west and south as the eye can see, from the mountains rising beside Winnimem and beyond, lies a constellation of other Wintu villages and village polities, like the one at Dekkas Creek. Many such villages flank rivers like the Sacramento, the Trinity, the Pit, and some of their tributaries. Speaking always a dialect of a single Wintu language, each group can communicate with the others, although diet and customs may vary.

With the gift of a common tongue, many neighboring tribelets are able to link through bonds of kinship or marriage, ceremony, and trade. From the people along Winnimem flow trade items such as salmon flour and pine nuts. The return flow from valley tribelets might bring roasted grasshoppers, acorns, or rabbit skins.

Neighboring peoples tend to be friendly, if somewhat suspicious, toward the people of Dekkas village, but short bursts of intertribelet warfare can occur. Battles have taken place between Wintu and Yana, Achomawi or Shasta groups; they are, however, a rarity. The Wintu are a peaceful people, and violence is usually restricted to sporadic disagreements like border disputes involving small-scale skirmishes.

Looking out across the mountainous terrain where Klamath, Coast and Cascade ranges rise from the floor of the great Central Valley, we begin to perceive a pattern — a patchwork of village territories strewn across the land, each complete in itself, sacred, and mindful of immediate, unmapped, geographic boundaries. We can imagine a delicate network of channels between village populations superimposed on this scheme of partitioned resources, marking exchanges both social and material among associations of familiar tribelet neighbors.

In many ways it is a simple system, this tiny Wintu village on Dekkas Creek, but it has proved a stable one. Experience through time has rendered the system effective; it has fostered survival and so it has itself survived.

A SOUND LIKE no other in the world pierces the taut silence within Dekkas, bringing a warm flood of relief to villagers who have managed only a fitful sleep this night.

It is the first scream of a newborn child — a baby boy that will grow to be a *wita*, a fine Wintu man whom the village can be proud of. His voice is full of terror and joy as he faces a spring night's cold breath; only moments ago all had been timeless warmth and the soothing beat of the mother's heart.

Inside a brush hut beyond the village edge, the mother comforts her child with soft words, then passes him without reluctance to the old women beside her. According to Wintu tradition, the boy-child must now be removed from the harmful supernatural forces that temporarily surround his mother following birth. During the next hours she will be massaged by experienced hands, and later she will cleanse herself with a steam bath, and, before she can receive her child again, have her abdomen bound with strips of buckskin.

The baby will be bathed several times each day, and when its umbilical cord has withered and fallen, someone may place it in a small basket and lay it beside the child's cradle. Later it will be hidden in a special place, or buried.

After one month of isolation and maternal dietary restrictions, the mother will return with her child to the family dwelling. A friend of the child's father will run around the hut carrying a cradle woven of hazel in a ritual welcoming the family's firstborn. Both parents will bathe before reuniting and, in the months that follow, will comply with other traditions to ensure their son's health and success.

The boy will be nursed for from two to four years and will not receive a name until he is old enough to understand its meaning and his lineage. Unlike adolescent boys of some neighboring tribes, he will not be required to undergo a rigorous initiation rite to attain manhood. Because the Wintu depend upon a rich oral literature to store and transmit knowledge, the boy will learn his way in life through exposure to the wisdom, both sacred and pragmatic, embodied in endless myths, songs, and folk tales; and with each new day, he will be reminded, like all Wintu, that the morning sun must never find him sleeping.

BY LOOKING TO the rock slide that descends from the high limestone cliffs just upstream from the junction of the McCloud and Pit Rivers, the people of Winnimem can predict the size of the spring runs of king salmon. If the slide is large, they are assured of good fishing in the coming months — a welcome influx of rich protein to the river-dwelling Wintu.

The Wintu are adept fishermen and employ a number of techniques to capture the hordes of four-year-old fish as they make the long journey from the sea to their spawning grounds. The man of the family sometimes builds his own salmon house, or *buki*, a simple scaffolding of poles, forbidden to womenfolk, which extends out from the edge of the river. He lies prone on the structure, the forward half of his body sometimes enclosed in a crudely woven willow blind, and waits — eyes alert and harpoon poised. At the precise instant, he thrusts the 15-foot-long double-pronged spear into the unwary salmon swimming below him. The prongs — capped with detachable tips, or toggles, attached by thongs to the main harpoon shaft — hold the struggling prey as the fisherman pulls it within reach.

From time to time the men of neighboring villages will join forces in a communal salmon drive that may last all night. Several of them stretch a net across a river, while others walk to a point upstream and ignite their torches. With flames held high, the torchbearers, in a line across the stream, wade or swim slowly down the river, herding masses of silvery salmon into the waiting net.

As in other cooperative efforts, the catch is distributed fairly — the leaders of each group awarding a portion to each male present. The fish are carried back to the villages to be baked in rock-lined pits, or sun-dried and pounded into a fine salmon flour, *dyee*, that will be stored in baskets or traded.

Coordinated hunting strategies are effective as well with deer. After scouting the bone-dry canyons along the Winnimem, several men may set up a temporary hunting camp from which they will systematically set snares for several days. A more dramatic method consists of actively driving deer by shouting and beating brush to herd them into the range of hidden bowmen, or over a sheer cliff. Sometimes hundreds of people participate. The accumulation of bones beneath Nop-in-tal-lus (Deer-fall-over) is testimony enough to the success of this hunting method.

A deer provides many things for the Wintu. Its flesh is roasted in hot ashes, wrapped in strips around cooking rocks, or steamed in baskets. Bones are often ground into flour. Equally important are the materials that the deer offers to the craftsman: tendons to give strength and elasticity to the sinew-backed bow; hooves for making sacred rattles used by adolescent girls; curved nasal-bone fish hooks; and hides which, when scraped and soaked in a preparation of deer brain, may become blankets, breechcloths, skirts, or moccasins.

Cooperative efforts — whether for salmon, deer, grasshoppers, or bears — enable the Wintu to tap some animal resources at a far lesser investment of energy than can the individual hunter-fisherman. A sense of generosity, revered by these villagers, assures that the spoils will be shared by all and, by adding to the communal food cache, will bring further stability to the Wintu economy.

Papil (Summer)

With the passing of moons, *oltipa*'s cool greenness gives way to *papil*, the months of scorching heat that ripen manzanita berries. People of Dekkas village no longer restrict themselves to the central settlement of bark dwellings beside the river. Semi-nomadic within their tribelet-territory, they make several food-gathering excursions within a radius of a few days' walk from Dekkas. They sleep in temporary summer huts of brush, following and harvesting the fruits of their familiar land.

For men there are the usual hunting and fishing endeavors. Women pursue the timeless task of picking acorns, seeds, and berries, and of digging for the earth's rich bounty of edible roots and tubers with sticks from mountain mahogany.

NAKED EXCEPT FOR a buckskin breechcloth, a shriveled man, old now and lame since birth, labors patiently in front of his brush hut. He is called Yorun-towi, Maker-of-Valuables, by the people of Dekkas. In one hand, protected by a strip of buckskin, he grasps a piece of black volcanic stone from Glass Mountain. With the other hand he uses an antler-bone tool to methodically flake away chips of the obsidian.

Yorun-towi's specialty lies in crafting the tools of the hunters and the fishermen — rope snares of iris leaf fibers or grapevine, salmon harpoons and willow-basket fish traps, fish nets, and quivers to hold arrows. With the skillful use of this equipment, individual men are able to bring in a steady supply of game for the village population.

In small creeks like Dekkas the Wintu employ an especially effective method of fishing. Soaproot is pulverized at the edge of a quiet pool. It diffuses into the water and stuns the fish, causing them to float to the surface where they are collected.

The flesh portion of the diet varies from one tribelet to the next, according to the nature of the habitat surrounding the settlement. The taking of life, however routine and necessary, is not regarded lightly by the Wintu. Not only must the hunter fast, pray, and obey sacred restrictions from time to time, but the very act of killing involves more than his willful decision to do so. It demands the consent, in some sense, of the prey. When a hunter pursues a deer without success, he does not agonize over his miscalculation, but states simply, "The deer does not want to die for me."

HER DARK FINGERS move deftly to an ancient rhythm as she sits in dusty silence beneath the midday sun. Lost in private thoughts penetrated only by earthly odors of acrid smoke and baking salmon, she weaves a basket with the easy assurance of a spider spinning its first web. A serving basket, or *kolom*, woven in the twined technique characteristic of Wintu basketmakers, it has for five weeks slowly grown from the coils of moist willow branchlets and pine root strips which lie about her. The willows have been stripped of bark and are used as a foundation or warp. The pine root has been roasted, split, and soaked before being twined against the willow rods.

The woman decorates the basket, its basic structure complete, with an intricate overlay of bear-grass and the split stems of maidenhair ferns. In this motif of white and black she weaves her final designs, depicting in an angular geometric pattern the flight of a flock of geese. Her steady hands seem to draw subtle strokes of grass and fern, giving expression to her thoughts.

The Wintu are skillful basketweavers, capable of transforming an assortment of local materials into utensils ranging from close-twined waterproof cooking baskets to coarse open-twined burden baskets and baby cradles. They make bottomless hopper baskets and flat disk-sifters used in processing acorns, spacious storage baskets for salmon and acorn and buckeye, and miniature charm baskets endowed with protective powers and given as payment to spiritual healers of the village.

Baskets reflect the value of their manufacturer's labor by being used as items of trade or as gifts. Baskets from neighboring tribes, including the coiled work of the Central Wintun and the twined basketry of the Shasta, sometimes pass through Wintu territories and find a use in Dekkas village.

Among men, the currency of exchange takes a different form, usually strings of clam-disk beads, or to a lesser extend *dentalia* shells or magnesite cylinders, all coming from regions far beyond Wintu borders.

Among other treasured objects are bear hides, elk skin armor, woodpecker scalps, food items, and various weapons and tools. These things appear in bartering transactions between villages. The value of a trade object depends on its availability, workmanship, and on the resources of the buyer.

In a land where food is relatively abundant, and climate not unduly harsh, the exchange of materials between tribelets is an exchange of surpluses. Fully capable of self-sufficiency, each tribelet offers limited excesses of raw materials or manufactured objects for others, expanding the diversity of the day-to-day experience in its miniature tribelet-world.

Imports are often distinctive, bearing faint auras of the distant peoples from which they came. To a village population whose worldview is deeply embedded in a single, fixed expanse of land, trade is one way in which its sensibilities are extended, its traditional patterns changed.

AS THE FIRST RAYS of a late summer sun strike Tlilemtconos (Place-where-the-acorns-are-dancing) on the west bank of the Sacramento River, an enormous black oak sways in the midst of a motionless forest. High in its uppermost branches a man grips a limb, shaking it wildly until its bounty of acorns falls dancing in showers on the ground.

Below him, laughing women and children scoop up the nuts. They remove the coarse caps with their teeth, and throw the rest over their shoulders into the burden baskets they wear on their backs.

Hours later the weary band will return to their summer camp a few miles away to begin the tedious work of making the acorns into the palatable food that above all others provides sustenance to the Wintu. The acorns are shelled with teeth or with a small stone hammer. The young women then pestle the kernels against a flat grinding rock — each woman sitting so that her legs support a bottomless hopper basket. Other women shake the freshly pounded acorn meal on basket-disks and separate the finer meal adhering to the woven ridges. The sifted meal is finally placed in sandpits and rinsed with water for several hours to leach away the bitter tannic acid which masks the acorn's underlying sweetness.

Distinctive in texture and taste, acornmeal dishes are prepared in various ways. From the finest meal, the Wintu make a soup — a thick and nutritious porridge, or mush — to be eaten with the fingers from serving baskets. The mixture is cooked by transferring heated cooking stones from the fireplace into a large basket of soup and stirring them rapidly with an oak paddle to prevent burning the basket. Coarser meal is moistened and wrapped in maple leaves, placed in a rock-lined hole, and covered with dirt. A fire is then built on top so the acorn bread can bake slowly in the pit-oven for several hours. Both soup and bread are staple foods, making available to the active Wintu the rich supplies of fat and protein stored within the acorns.

THE WINTU POSSESS an intimate knowledge of the local flora, as they must to survive in an environment where foods are made from many kinds of widely scattered plants. They know each plant — when it ripens, where it can be found, how it can best be used — in order to efficiently harvest nature's bounty thoughout the year. They will at various times eat the starchy underground stems of *brodiaeas*, the bulbs of wild onions and lilies, the leaf stalks of wild rhubarb, and saxifrage, mushrooms, wild grapes, elderberries, chokecherries, blackberries, hazelnuts, pine nuts, grass seeds, and countless others.

The Wintu are but a small elemental part in the vast scheme of forest and river. Except for their occasional use of fire to clear brush or drive game, the survival activities of the small village populations scarcely alter the face of the land. The Wintu are generalists, adapting quickly to the change in season, committed to a tediously slow process of gathering foods that tend to be small and scattered, sharing or storing surpluses as they occur.

Kaidono (Autumn)

The parching winds of late summer, the ripe acorns on the oaks and the fall runs of salmon speak of the coming of *kaidono*.

For these people who see in all aspects of their surroundings the vital forces of the world of spirits, it can be said that their lives are permeated with the sacred and the supernatural throughout the year. A man visits a sacred rock (*sawal*) near his village to pray for luck in tomorrow's deer hunt; he carries with him a charm stone to ensure success. An old woman, sleepless for the aches and pains of her age, chews a piece of dried angelica root, called *chewachi*, to ease her suffering and to protect her from the suckerfish spirit. A child, asleep in the family hut, enters the realm of spirits — vividly experiencing a conversation with a mountain spirit that lives in the foothills above Winnimem. His dreams will be accepted as a portion of a reality in which wakefulness is only a small part. Not to be scorned, such visions are a direct communication with innumerable spirits including ancestors, springs, animals, and objects that sometimes offer prophetic pieces of the puzzle that is every man's life.

For the most part the spiritual life of the Wintu remains an individual matter. They live without the elaborate costumes, dances and ritualized ceremonies practiced by wealthier tribes in the heart of the Sacramento Valley. There are no first-salmon ceremonies or wealth-display dances like those practiced by the inhabitants of the lower Klamath River and the coastal areas to the northwest. Instead, the Wintu look to their most visible ceremonial displays — the *batlastconos*, or adolescent girl's initiation dance, and *thabitconos*, or shaman's initiation dance, as well as the regular healing séances of the shaman who cures the sick with the help of spirit aids. There are frequently carefree feasts with friendly village neighbors to share in games, songs and dances.

IN THE MIDST OF swirling, shouting dancers stands a young girl of fourteen years, dressed in a fringed skirt of doeskin and strings of olivella shell beads; beside her is a still younger girl serving as her attendant. Pallid and trembling slightly, the older girl, called Bulidakismet, or Snow-mountain-woman, struggles to maintain the proud and erect posture that befits the daughter of the Dekkas village headman. Betraying for an instant her exhaustion, she leans heavily on the striped staff topped with a cluster of tiny deer hooves, closes her eyes, and in a quick flood of memories recalls the long weeks of her ordeal.

From the time of her first menstrual period some weeks ago, she has been required, according to Wintu tradition, to live in seclusion, eating and sleeping in the brush hut at the edge of the village, and avoiding contact with most people during this time of spiritual potency. For the first five days she was not allowed to sleep; instead, with the aid of her deer hoof rattle, she maintained a continuous vigil against evil spirits that lurked nearby. A meatless diet of acorn soup was brought to her each day, and village elders sometimes visited to offer advice concerning her duties as a mature woman of Dekkas village.

Now, at the onset of her second menses, Bulidakismet is to receive the formal public recognition that is her due as a member of an important family. Her father has sent messengers to neighboring villages to invite them to the *batlastconos*. Now there follows a special time, with feasting, storytelling, games and singing in her honor. There are dances held each day, and, with these, the final hours of her *batlastconos* drawing near, the dance called *sedemtconos*, Coyote-dance, begins. An array of dancers — the women in two lines, the men in a central circle — raise bows and arrows or bunches of feathers in the air as they hop to the wild rhythm of elderberry split-stick rattles slapping against hands.

Besides the endless baskets of acorn soup, venison and salmon, there has been time for sweat baths in the earth lodge, gambling, and friendly conversation over pipes of tobacco concerning this season's hunting, news from a more distant tribelet, the possibility of a daughter's marriage to a Dekkas man. Even with the last dances, the ceremony nearing an end, some people will be reluctant to leave friends and relatives, and will linger for still another round of hand games. Among them will be men who have in the last few days won a small fortune in clam-disk beads, and who will offer losers a last opportunity to reclaim their losses. Men spend hours in anticipation of gambling competitions — honing their skills, fasting, or visiting sacred places that can bring luck to a gambler.

When the last guests have finally departed for home, Bulidakismet will be free to return to her family dwelling. She will bathe and discard her ceremonial clothing before entering the hut where she will rest, weary but somehow strengthened by her experience. She will relive it, in part, with each month's routine isolation in the menstrual hut. This routine is shared by all Wintu women; they must obey the solemn demands on womanhood or face spiritual retribution.

If her life follows the established path of most women, Bulidakismet will one day take up residence with an eligible man of an adjacent village after he has presented appropriate gifts to her family. The two families may then sanction the marriage by exchanging gifts of equivalent value, before the couple move in with one family or establish their own home. The bond may last a lifetime, or considerably less, depending on its strength and on the events that follow. From it will come the seed for future generations of Wintu, whose lives, like Bulidakismet's, will take root in the rich and stable soil of ancient and unquestioned traditions.

Pomisin (Winter)

In the warmth of the family hut, a dead boy's father prepares the body for burial. He dresses it in buckskin clothing; then, forcing it into a hunched posture with elbows between the knees, and hands against the face, wraps it in the hide of a bear. He binds it tightly with deer sinew and prepares to carry the small bundle through a special opening made in the rear wall of the dwelling.

Outside in the village graveyard two old women dig patiently with sticks in a drizzling rain. Not a word is spoken as they strain to lift baskets bulging with dirt from the round grave-pit. They have excavated many graves over the years, but it is especially painful to see a child's frail body laid in the ground.

The burial will draw visitors from several villages, and the boy's younger brothers will be swung over the open grave to ward off other deaths in the family. The body will be covered with pine bark and earth, and bereaved relatives will rhythmically stamp on the grave as they cry or sing.

If an adult had died, the burial would have been followed in time by the destruction of the deceased's personal property. Weapons and tools would be shattered and other belongings burned so as not to attract or offend the departing spirit.

The boy's only possession, a small yew bow, lies beside him in the grave. At his right side rests a basket filled with acorn water to quench his thirst during his soul's inevitable journey — north to Mount Shasta and the Milky Way, then south and east to a fertile plain to join with other dead Wintu's souls.

After the funeral, many villagers will purify themselves in the smoke of an oak-wood fire or in the heat of the village sweat house. There, soothed by eddies of warm air, a man may offer words of advice and consolation to the dead boy's grieving brothers. He will warn them never to utter their brother's name lest they disturb his spirit, which still lingers nearby. Perhaps the man will comfort the boys, too, by relating the legend of how Coyote brought death to the Wintu, and was its first victim.

TWO DISTANT FIGURES stand, shivering, beside the snow-covered mounds of the Dekkas village cemetery. They talk in hoarse phrases, their voices muffled by the fresh drifts of snow, and sometimes pause to tilt their heads back to draw smoke from pipes of tobacco. One of them leans over a small grave and sprinkles red water from a basket with her hand.

At the other end of the village, inside the snow-covered dome of the earth lodge, a group of men and boys is warmed by the hot, smokeless fire of manzanita wood. The bodies of all of them are drenched with salty sweat, and the air is heavy with odors. From time to time, someone from the group crawls out the overhead smoke hole and sprints over the snow and into the icy waters of Winnimem.

During these months of restrictive cold, the Wintu here will subsist on the huge baskets of acorns, salmon flour, dried venison, and other foods stored in previous seasons. Small game and deer sometimes supplement the diet. Famine is unlikely; however, should severe weather linger longer than usual, people will be forced to hunt more often, and possibly resort to eating the sweet inner bark of yellow pines. For the most part, their work activities in the Dekkas settlement will consist of gathering dead wood for fires and rolling snow into great snowballs to clear village paths.

There will be long hours in the sweat house, where, with past sorrows forgotten, men will gamble and sing and joke away the winter hours.

Even now young Wintu boys listen in silent amazement to men's tales of strange events in the world of Wintu spirits. One man reaches into a pouch to fill his pipe with wild tobacco as he tells a short tale. He tells of a place up the river, along the west bank of Winnimen called Kokospom (Flea-ground). Once, long ago, two fleas were traveling the banks of the river, one heading upstream, the other down. Each was lost in his own thoughts and did not see the other approaching. Suddenly, precisely at the spot called Kokospom, the two daydreaming fleas crashed into each other. Their heads hit with a force that broke both of the fleas' backs.

"That is why," exclaims the old man, "today all fleas have humped backs!"

Before a boy can ask a question, another man leans forward, and, gazing into the fire says, "Long before this world existed, there was Mosquito who sucked the blood out of people. He sucked so much that his stomach was full and he could hardly fly. He went up above and gave some of the blood to Thunder. Thunder said, "This is fine. Where did you get it?" Mosquito answered, "I got it from the biggest trees that grow high up on the mountains. The little trees haven't any blood." That is why Thunder, and Lightning, who works with him, always strike the big trees up on the mountains. They are looking for blood. If Mosquito had told him the truth, Thunder and Lightning would get people."

A BOY LAUGHS SOFTLY as he thinks of the strange tales, then leans back on his bed of evergreen boughs and drifts into sleep. Once again a spark of the rich, unwritten oral literature of the Wintu has leaped from one generation to the next and for a time found new life.

1977

Flora, Shaman of the Wintu

IN THE YEARS before their world was rocked to the core by nineteenth-century European invasion, the Wintu traditionally initiated shamans into their roles as tribal healers and spiritual guides through elaborate communal ceremonies.

The rite began in the evening, with the shamans and shaman candidates dancing naked around a manzanita-wood fire and singing to invoke the spirits. A whistling sound above the smoke hole of the earth lodge signaled the arrival of the spirits. If a spirit found a candidate acceptable, it entered his body. When this happened, the novice's behavior became frenzied: his body jerked convulsively, saliva poured from his lips, and sometimes blood flowed from his mouth and nostrils. Finally, he fell to the ground and was carried by the older shamans to one side of the lodge where he was carefully watched over and sung for. Each man or woman thus called upon by the spirits was then recognized by his tribesmen as an individual who possessed a potent rapport with the world of spirits — a world that the shaman would reenter during the trance of healing séances countless times in the future.

Today, memory of language or traditions is rare among the Wintu people and only one acknowledged shaman remains — an elderly, vibrant woman named Flora Jones. Living on a modest ranch located within former tribal boundaries, Flora — or Pui-lu-li-met, Eastern-Flower-Woman, her Wintu name — remains an active practitioner of the ancient healing arts of her people. She still conducts shamanistic séances and administers traditional herbal medicines to patients of her own neighboring tribes.

I first met Flora in 1972. As our friendship grew, we would spend hours discussing her healing activities, sometimes visiting sacred doctoring places along the McCloud River near her home. In the beginning Flora voiced misgivings about my questions, saying at one point, "Of course the spirits that I doctor don't like the white people." Then she would add, "But I have been praying for you since I knew you. I tell the spirits that you are my child and they must accept you. And that I have looked you over thoroughly and you mean no harm."

Although her reluctance to discuss certain matters proved frustrating, it came as no surprise. Her reticence had been voiced by another Wintu shaman in a dream song recorded by ethnographers more than 40 years ago:

When Red Cane [white man] comes
We Wintu shall forget our songs.

Because her people's culture has undergone catastrophic changes in the past century, Flora's practices cannot be viewed as a crystalline record of Wintu shamanistic arts; rather they represent a mix of modern practices and traditional spiritual doctoring. Nonetheless, the traditional elements remain, and the story bears a special importance at this time, for there is no apparent successor to her sacred office. As she herself knows, Flora may well turn out to be the last of the Wintu shamans.

Flora calls herself an Indian doctor. Most anthropologists, however, would refer to her as a shaman since the definitive feature of a shaman is the capacity to go into trance — the ecstatic state in which the supernatural world of tribal spirits is directly experienced. Shamans often describe the ecstatic trance as a journey of the soul. Wintu accounts suggest a somewhat more passive role for the shaman. Flora explains, "The spirits use me for just a tool — something to get into to tell the people what is going to happen and what they can do." Here a shaman's powers of prophecy and clairvoyance seem to arise not from the wanderings of the soul, but from the omniscience of certain mobile helping spirits, as well as the omnipresent guiding force of the Wintu creator-god, Olelbis.

Wintu helping spirits take many forms: they may be spiritual forces residing in sacred places or celestial bodies, the spirits of animals, or the souls of departed relatives, among others. Flora's own pantheon of spirit helpers contains representatives of each, acquired through long years of doctoring experience, and each is endowed with its own personality and usefulness.

She describes the suckerfish spirit as a trickster, a mischievous figure capable of communicating in many tongues and one not to be trusted during a trance. Another, the black wolf spirit, projects a hostile character and appears most often during times of deep distress. Others — a star spirit, a moon spirit, a mountain spirit, the soul of an ancestral shaman — may appear to her during a séance as well. When several spirits enter simultaneously, engaging in animated discussion during a trance, the mountain spirit acts as leader and spokesman.

Occasionally, I would ask Flora to list her helping spirits, only to discover that the "list" grew longer with each attempt as she recalled still other sacred mountains or springs or rock formations. The spiritual world of the Wintu, it soon became apparent, is not easily circumscribed. A suggestion that she try to sketch one of her spirits drew a burst of friendly laughter. White people did not seem to understand, she told me, that the spirits could not be drawn, for their appearance was fleeting, beyond description.

Although Flora became a shaman at a time when the ceremony of the communal spirit-quest, or "doctor's dance," described above had been lost to the Wintu, her initiation experiences revealed features common to traditional shamanism in both her own and other cultures.

Flora, an adolescent at the time, was attending a federal school in San Francisco when she first experienced the vivid dreams that eventually led her home to become a shaman. These dreams confused and terrified her, but they also challenged her to resolve the dilemma of living in a world that was to her both Wintu and white.

From the beginning her life was a paradox. Born in 1909 of racially mixed parentage (her great-grandfather was the last great chief of the McCloud River Wintu), she was baptized a Methodist — yet accepted the reality of childhood encounters with tribal animal spirits. Although formally educated in government schools, she was reared by a mother who spoke Wintu and often relied upon Wintu shamans to maintain her daughter's health. As a young girl,

Flora was lured by fantasies of a career as a Hollywood actress — yet she shot her first deer at the age of 10 and, following tribal tradition, abstained from the eating of this first flesh.

At the age of 17 came the climactic event of her calling — her first trance. She was engrossed in a card game with friends when, without warning, her ears were filled with a ringing sound and a burning pain.

"It was like a hot bullet going through my ear," she recalled. "The pain went through me and I passed out for four days." Years later she viewed this traumatic experience as her first ecstatic encounter with a helping spirit — the star spirit.

She awoke from her prolonged unconsciousness singing. With her were four older Wintu shamans, who in a manner reminiscent of the traditional initiation ceremony, sang and cared for her, administering medicine and later taking her to sacred places to pray. In the days that followed, her helping spirit taught her doctoring songs and tutored her in the healing arts of the shaman. Finding her acceptable in their own trances, the elder shamans established her as a shaman in her own right, although she did not begin doctoring until years later. By singing songs, visiting sacred places, and entering trances, Flora found that a new calm had settled over her. No longer, as in years past, was she beset by mental turmoil and sickness, by the contradictions of her upbringing, and by self-doubt; no longer were "the spirits taking her to the mountains."

Some of the most illuminating conversations I have had with Flora were held at a sacred place where she often carried out séances. Located in a shady stand of black oak trees, it was, she said, the former site of an earth lodge, or *hlut*, where Wintu doctors had once healed the sick. At such times, cheered by the familiar surroundings, Flora talked at length of her life as a doctor, her voice sometimes low and resonant as she spoke in careful phrases, sometimes rising to an excited pitch as she described a particular dream or premonition.

She told me of the five kinds of Wintu doctors. Each had his own area of expertise but lacked the high degree of specialization attained by shamans of certain neighboring tribes of California Indians. The first two, the dreamers and the singers, were shamans possessing powers of clairvoyance not associated with curing. Tracer doctors were adept at locating lost objects or souls. And sucking and healing doctors were primarily active in the realm of healing.

Although each of the five types worked through a trancelike state, the sucking doctors were traditionally the most powerful and highly regarded by tribesmen. They cured by extracting minute pathogenic spirit-missiles, or "pains." They removed the pains by sucking the afflicted area of a patient's body while in a trance — a procedure that must have positively affected local blood flow as well.

Flora considers herself a healing doctor, diagnosing illnesses with her hands during an ecstatic healing séance. But like many shamans she has displayed elements of more than one specialty. For example, as a healing shaman, she has sucked pains from certain patients, but like a tracer doctor, she predicted the location of a missing corpse on a nearby river.

Her séances usually take place in the company of a traditional assistant, or "interpreter," the patient, and various friends and family members. Despite her willingness for me to do so, I have not yet witnessed one of Flora's doctoring ceremonies and must rely on her accounts, as well as tape recordings of some of her séances, for my descriptions.

Oddly enough, the tapes were Flora's own personal recordings, for she had on occasion been forced, by necessity, to use a tape recorder in place of the elderly man who usually served as her interpreter. By later replaying the tapes she was able to penetrate the amnesia that inevitably followed her trances — the tapes acting as a surrogate interpreter, revealing the supernatural dialogue that had taken place. At first the presence of this mechanical aid apparently offended her spirit-helpers, but under the circumstances they proved in time quite willing to tolerate it.

Once assembled, the participants in a séance join the shaman in summoning the helping spirits by singing the appropriate Wintu songs. (One such song, according to Flora, describes a shaman's search for a path taken by the souls of the dead to heaven and his or her conversations with trees and flowers along the way.) Periodically, the shaman accepts a pipe of tobacco from the interpreter, inhales deeply, then resumes singing. Traditionally, shamans used a species of wild tobacco (*Nicotiana bigelovii*) to enter a trance; today, commercial tobacco must suffice.

The arrival of the spirit-helpers is marked by the shaman's sharp inhalation; the group falls silent as the spirits begin to speak through the shaman's voice, offering advice or prophecies and sometimes quarreling or joking among themselves.

The shaman drinks clear water, then a solution of acorn water from a small container — an offering to the helping spirits. With the diagnostic powers of the spirit-helpers acting through her hands, she begins to move her fingers carefully across the patient's body, sensing unseen internal injuries or abnormalities.

"I feel for the sores, the aches, and the pains," explained Flora. "When I put my hand over the body I can feel every little muscle and every little vein. I can feel the soreness. It hurts me. If they have heart trouble, my heart just beats. Any place they are hurting, I hurt. I become a part of their body."

If the spirits detect the source of the sickness, they offer prescriptions for the patient's care, speaking through the shaman. Remedies may be offered for both physical and psychological ills and often include traditional herbal medicines, which Flora meticulously collects and stores.

Upon awakening from the trance, the shaman learns of the spiritual diagnosis from the interpreter and proceeds to treat the patient accordingly. In the past, a patient's family would present the shaman with a tiny, handmade basket of characteristic striped design following a successful treatment. Now the transaction usually involves money and takes place regardless of the outcome of the therapy.

Only in the most dire cases, when a patient is unconscious and death imminent, does Flora perform the Soul Dance — a final attempt to recapture the victim's wandering soul. This ritual is performed at midnight. As the shaman dances to the rhythmic beating of sticks she waves a staff bearing a miniature basket in which to catch the lost soul. If the shaman's search is successful, the soul is brought back to its owner and placed over his heart, its natural resting place.

When not in use, the staff and other shamanistic regalia are kept out of sight, often in a special hiding place, and treated with great respect. They include sacred "yellow hammer" feathers, which can be used to "pull evil dreams" from a person, and a wooden pipe possessing spiritual powers that can be consulted for advice. In addition, Wintu shamans have included among their doctoring paraphernalia rattlesnake rattles, rattles made from cocoons, and the hollow leg bones of birds (to contain extracted pains).

At one time, the mere presence of a white man at a séance was sufficient to disturb Flora's trance. Gradually, and with apparent reluctance, her helping spirits have come to accept and diagnose white patients as well as Indians.

"Nowadays," Flora stated solemnly, "we have to live almost half in the white man's way and half in my own way. I will never forget my own way, but I have to steer myself with the white ways. The spirits understand that... I was told by the spiritual one time when I was doctoring that I must help my white brothers as well, if I am to be a thoroughly spiritual doctor."

Treatment of whites, however, seems to demand different techniques. Diagnosis takes place without an interpreter and involves only a brief trance. Like some sort of transient X-ray image, the trance apparently permits Flora to visually scan the patient's internal organs for the source of his or her discomforts.

Wintu shamans traditionally ascribed three possible causes to most diseases. The first and most commonly encountered were the *yapaitu dokos* — minute disease-objects, or pains, which were said to be sent by an offended spirit or perhaps a sorcerer and which were often thought to be detectable in the blood. The origins of such an explanation for disease are obscure, but it is interesting that the eminent naturalist C. Hart Merriam reported that Wintu shamans once prepared and surreptitiously threw tiny poison darts at selected victims. Diseases are also thought to arise through spirit possession, especially in situations involving the violation of a particular taboo. Or a person's soul, it is said, may be simply lost — released during a period of unconsciousness or stolen by an evil spirit.

Flora herself attributes most diseases to some form of self-inflicted stress, emotional or physical, and seldom mentions such traditional pathology. Although she frequently resorts to the use of medicinal herbs, much of her treatment lies in preventive medicine — giving advice on personal attitudes, care of the body, and proper nutrition, for example. She is highly critical of processed foods and refuses to drink tap water, preferring instead to journey several miles to fill containers with water from sacred springs. Her diet, too, is restricted, partly in deference to traditional food taboos established by her spirit-helpers. She maintains that the discipline imposed by such a regimen, along with advancing age, tends to strengthen the powers of a shaman.

Some diseases, unless detected at an early stage, are considered beyond the healing powers of a shaman. Most of these are categorized as "white man's diseases" — ulcers, cirrhosis of the liver, and certain cancers, for instance. Actually, according to early-twentieth-century anthropological studies, a host of other diseases were also introduced to California Indian populations by newly arrived Europeans. These include smallpox, tuberculosis, cholera, typhoid, scarlet fever, measles, whooping cough, syphilis, and gonorrhea. This adds deadly irony to the Wintu's choice of the words *yapaitu wintun* — sometimes translated "poison people" — to describe the white intruders who were often simultaneously bearers of deadly infectious diseases as well.

Prophetic dreams occupy a central place in Flora's healing activities and have been an important factor in the shaping of her worldview. She has been convinced through her own dreams, for example, of the impending destruction of the world — a feeling reflected in the following statement: "The Indians are going to get down very low [in numbers], but this world is getting so near to the end. The white people are getting too close to the moon and the sun. The white people are causing the world to end."

The prophecy of a world corrupted by whites and on the verge of catastrophe is not a novel one among Wintu shamans. Faced with the painful erosion of their culture under the forces of a relentlessly dominant colonial society, the Wintu have openly entertained such thoughts since at least 1870, when the revivalistic Ghost Dance Movement, with its dreams of the dead and its predictions of the world's end, swept through their territory.

In response to some of her dreams, Flora has recently joined others in trying to generate a spiritual renaissance among people of her own and other tribes. Speaking of her sacred obligations as a shaman, she told me, "This is what the spirit tells me: get my people together. Get them to believe, because if you don't, they are going to go wild. They are going to kill one another. Whoever has sacred places must wake them up, the same as I am doing here — to keep my old world within my heart and with the spiritual. For them to help me to help my people."

In an attempt to "wake up" certain Wintu sacred places and

revitalize her doctoring activities, Flora in 1973 approached the local U.S. Forest Service administration, which now controls vast tracts of traditional Wintu lands, for official permission to carry out doctoring ceremonies on public lands. The Forest Service, initially baffled by this novel request, eventually conceded — but not without characteristic bureaucratic caution. The historic document presented to Flora read: "by means of this letter you are hereby authorized to utilize approximately 1 acre of land . . . to carry out your Indian doctor practices. This does not authorize the violation of any County, State or Federal Laws."

Flora has also periodically conducted a special prayer ceremony on a mountaintop near her home. Obeying dreams in which helping spirits appeared and provided instructions, she entered a trance and, waving her shaman's staff toward the four directions of the compass, offered prayers and advice from the spirits that speak through her.

Once, when I was photographing Flora at one of her sacred places, she startled me by asking if I had sensed the spirits around me. They had been examining me, she said, and in the process she had seen my face disappear for an instant as I sighted through the camera. "Were you dormant?" she asked intently.

From the beginning, my relationship with Flora was based upon both my intense curiosity and my innate skepticism. The latter, I knew, blinded me to much of what she earnestly tried to communicate to me. In the end, I always seemed to feel compelled to reduce her shaman's world to what I believed must be its underlying scientific roots. Yes, I was forced to admit, I was dormant; I had *not* sensed the spirits about me.

Many attempts have been made in the literature of anthropology to suggest a physiological or psychopathic basis to shamanism. Inevitably, this attitude seems offensive or foolish to those Indian people who still believe strongly in the old ways. Some researchers have pointed to a possible constitutional predisposition among certain individuals to become shamans. Others have emphasized, in some cases, the role of hallucinogenic drugs, while still others have proposed environmental factors such as sensory deprivation (among Arctic peoples, for example), or medical ones such as brain lesions and epilepsy.

Epilepsy in particular has offered intriguing possibilities in certain cases. It has been at least implicated in studies of shamanism from a wide variety of cultures, with reports of seizures and other characteristic symptoms of epilepsies. Such features are discernible, at times, in Wintu shamanism as well. Convulsive seizures seem to frequently accompany the initiation rites of shamans. And there are the perplexing and often strikingly uniform reports by shamans of sounds that often precede their trances. "Whizzing," "buzzing," or "ringing" in the ears, sometimes accompanied by abdominal pains, appear repeatedly in their descriptions of the auralike premonition of a trance.

I once asked Flora directly if the sensations she experienced while entering a trance might not be something akin to what an ordinary epileptic feels. The term *epileptic* did not surprise her; she often injected modern medical terms into our conversations. Her answer was equally direct: "It does not necessarily have to work through epilepsy." Then pausing thoughtfully, she added, "The spiritual acts in *many* ways."

Those ways remain, I am convinced, largely the province of the shaman's extraordinarily rich internal world — essentially invisible to an analytical mentality. That the shaman is possessed of considerable strength, stamina, and personal discipline, as well as genuine healing power — whether in the realm of herbal medicine, psychosomatic medicine, or his or her own psychotherapy — there is, I think, little doubt.

Perhaps those of us who fail to fully comprehend the time-tested techniques of Wintu ecstatic healing are, as Flora once suggested to me, simply suffering from a sort of a spiritual dormancy. Such people, she maintains, are deserving of her sorrow, for they live out their lives "with clouds in front of them."

1975

Weavers of Wood: The Ancient Art of Haida Basketry

IT WAS RAVEN, the Haida say, who created their ancestral home — the Queen Charlotte Islands, an archipelago off the coast of British Columbia. In a mythical time, when nothing existed in the world but an endless expanse of sky and sea without shore, Raven took magical stones from a powerful being beneath the ocean and spat them from his beak. The fragments of rock grew to form a cluster of beautiful islands in the sea, each covered with magnificent forests of spruce and cedar trees. This same hero and trickster, Raven, later brought the Haida people gifts of daylight and fire and fresh water — and also the skill, some say, to weave the first Haida baskets.

Early Europeans arriving in the resource-rich Northwest Coast area found the Haida living on the archipelago, as well as on the southern end of Prince of Wales Island in Alaskan waters. Along with their Tlingit and Tsimshian neighbors, the Haida have long distinguished themselves in the realm of Northwest Coast Indian art. Haida culture has given birth to generations of master woodcarvers and painters and, in historic times, to talented argillite carvers and silversmiths. Traditionally these artists have been men, and today several have an international reputation. In contrast, the baskets produced by generations of Haida women have not always been accorded the same artistic status, although they were recognized, in the words of anthropologist Erna Gunther, as "a fine utilitarian ware that was distinguished by the excellence of its craftsmanship." This reputation is shared with other Northwest Coast Indian peoples, including the adjacent Tlingit and Tsimshian. (Tlingit women also wove the famed Chilkat blankets of cedar bark and mountain-goat hair, which have been compared to the finest ancient textiles of Peru and Mexico.)

In the absence of pottery making, basketry evolved into a complex craft that was a major component of Northwest Coast aboriginal art. Proficiency in basketry was demanded of each female as she matured within the matrilineal Haida society. Males might assist a basket weaver in some way, perhaps in the gathering of materials, but generally no Haida man would want to weave a basket. And women were usually forbidden to use the totemic motifs of animals, humans, and spirits that adorned the carvings of Haida men.

A Haida basket is a tree transformed. This one realizes while watching, for the first time, a Haida weaver, deep within a shadowy Pacific coastal rain forest, straining at the end of a great ribbon of bark that she is pulling from the trunk of a tall western red cedar. A modern weaver does this by first making a horizontal cut at the base of the tree with her hatchet. Then with vigorous upward movements of her arms, she rips a broad strip of bark from the tree until it is far above her head. Once she has jerked the bark from the tree, she uses a knife to peel away the light inner bark lining in long strips, which are tied into bundles. The weaver takes care to harvest only a limited quantity of cedar bark from any one tree, so that it will heal, and sometimes pauses to whisper a friendly assurance or greeting to the generous tree. After the inner bark has seasoned for one year, it is separated into narrow fibers suitable for weaving.

The finest and most admired examples of Haida basketry are those constructed from the slender roots of the Sitka spruce. Among the Haida, there may be no more than a dozen basket weavers alive today, scattered among villages and towns in southeastern Alaska, the Queen Charlotte Islands, and mainland British Columbia, who are expert at working with this material. Two of the most respected are master weaver Selina Peratrovich, born in

1889 in the Queen Charlotte village of Masset, and her daughter Delores Churchill; both now live in the island town of Ketchikan, Alaska. As major native Alaskan artists, these two women have fostered a renewed interest in Haida basketry art among Haidas and non-Haidas by exhibiting their work, by serving as consultants to major museums, and by offering expert instruction to students. Recently, the two weavers were the subjects of a U.S. National Endowment of the Arts videotape study in which I participated — another attempt to foster and preserve this timeless art.

In spring or early summer Haida women have for centuries ventured into the towering gray-green stands of spruce that embroider the sandy shorelines of the Queen Charlotte Islands to gather roots to make their baskets. Beneath the familiar undergrowth of salal and thick moss, they would seek spruce roots or proper flexibility and dimension — no larger than one's little finger in diameter. Traditionally, they located such roots by digging a trench perpendicular to several roots with a fire-hardened digging stick. Today, garden tools and a pair of work gloves often suffice.

In order to prepare them for weaving, coils of spruce root are carefully roasted on the embers of an outdoor fire. A length of root is lightly charred for a few minutes, then quickly uncoiled and threaded through a vertical slit in a wooden stake anchored in the ground. Pulling the root firmly through the slit removes the leathery bark by abrasion, leaving in hand a glistening aromatic cord the color of straw. Says Selina Peratrovich, "You put it in the fire — but not in the red coals. You steam it; then you take the skin off." For her, the root is sufficiently roasted when it begins to produce a distinctive hissing sound; her daughter insists, with a trace of a smile, that she has managed to successfully prepare roots without ever having heard such a convenient cue.

Each root is then split, using a knife to start the cut; in the past, the sharp-edged shell of a mussel was the tool of choice. Gripped by fingers and teeth, the strands are then pulled apart. For a thick root, the woman lays several cuts in a manner that guarantees the piece will be fully utilized. The polished strips that formed the outer root are reserved for the most visible part of a basket; stout inner strips are employed for a basket's framework; the soft central pith is discarded.

The basic architecture of Northwest Coast baskets follows either of two primary designs. In a "coiled" basket, the foundation consists of a spiraling bundle of fibers, the coils of which are sewn together by loops of fiber. An awl or needle is used as a tool in the process. A "woven" basket, on the other hand, requires no sewing tool: a flexible fiber — the weft — is twisted in and out of a series of stationary fibers — the warp. The technique can be anything from a quickly executed "plain" weave, in which the alternating undulations of strands of similar width form a checkered pattern, to an elaborate braiding of multiple weft strands that creates rows of decorative texture.

The Haida traditionally produced only woven baskets. Checker techniques were employed largely for common household ware, especially vessels used for hauling and storing, usually made of cedar bark. Cedar bark also found its way into mats that served as bedding, rugs, or room dividers; into capes worn to shed the incessant coastal rains; and possibly before and certainly following European contact, into simple sails for some ocean-traveling Haida canoes. Twining — in which two or more weft strands are twisted around each other as they are woven into the warp — dominated Haida work, however, and represented its highest refinement. Tightly twined spruce root formed the typical Haida cylindrical storage basket, and the ceremonial dance hat.

The intricate variations of twining were cataloged in detail for Tlingit basketry in a classic 1903 publication by George T. Emmons, a naval lieutenant with a keen interest in Northwest Coast artifacts. Emmons wrote that the Haida shared many of the specialized weaving techniques employed by the Tlingit and that in Haida work generally, "the ordinary Tlingit weave in spruce-root obtains, although the ornamentation differs." Emmons lists five separate styles to twined weave, ranging from those that create a sievelike fabric to others that yield decorative surface textures. The Haida used the former, for instance, in their sturdy, undecorated openwork baskets for gathering razor clams, edible seaweeds, and berries. The latter might be a rigid ropelike weave, made of three weft strands instead of the usual two, used to reinforce basket stress zones, such as the junction between bottom and wall or the perimeter of the lip.

A Haida basket weaver begins her work with several warp fibers, which she bends and binds into a whorl; this basket "start," as it is called, is the embryonic base of the vessel and serves as a first scaffolding upon which two or more weft strands are twined. The bottom of the basket is completed first, then the work is inverted so that the weaving progresses from the base downward to the lip of the basket. This appears to be a distinctively Haida custom. Traditionally it was done by suspending the basket from an overhead cord; a modern weaver achieves the same result by tightly constructing the walls of her inverted basket over the frame of an empty bottle or can.

The strands of fiber are soaked before being used. Using moistened thumb and forefinger, and a simple weight to anchor the start at first, the weaver performs the basic twined weave by enclosing, in turn, each of the radiating stationary rays of warp with a pair of weft strands, crossing the latter between warps. By pressing each row of spiraling weft snugly against the preceding one, the weaver produces the beaded, vertically ridged texture typical of so many Northwest Coast spruce root baskets.

Additional lengths of spruce root are spliced into place, as extensions to the first warp fibers, by the encircling strips of weft, so that the basket skeleton actually grows with the piece. Like a potter at a wheel, the Haida weaver can effectively shape her emerging vessel by varying the number of warp rays as she weaves: by increasing their number, she gives her basket an outward flare; by reducing their number, she pulls the shape in; by bending all the warp rays at a right angle, she achieves a transition from basket base to basket wall. The Haida basket weaver concludes her piece with one of several possible border weaves in the final rows of weft, anchoring and reinforcing the exposed edge of the basket vessel. In its simplest form, found especially in decorative ware, a border weave is nothing more than a last spiral of twined weave, following which the end of the weft is secured with a terminal knot and the protruding warp strands are trimmed. In other cases, the Haida might employ the more sophisticated three-strand twining, which they are so adept at.

Haida baskets are characterized in part by the simplicity of their ornamentation, often nothing more than the patterns of raised ridges created by a colorless decorative weave or by the simple substitution of a colored weft strand for the usual undyed one.

When color did appear in the Haida basket, it was traditionally the stark black of local muds. Other, subdued hues — the pale yellow of a lichen, the green of copper ore, the reds and purples of berries and barks — also were eventually used. In historic times, Haida weavers began to decorate with a colored fiber wrapped around the outside loops of weft, an overlay technique called false embroidery, of which the Tlingit were masters.

According to Delores Churchill, the Haida had a repertoire of named basketry designs, as did their Tlingit neighbors. Often the names referred to natural objects or processes that to the non-Haida only remotely correspond to their geometric form; today the meaning of some names has even been entirely lost. Delores Churchill lists several "very old" Haida basket designs, among them, "fishbone," "cresting wave," and "spider web."

The Haida basketry hat — an elaborate broad-rimmed headpiece constructed entirely of spruce-root — is perhaps the single most elegant manifestation of Haida fiber art. With a crown sometimes woven with variations of the three-twined weave, with a brim woven in the basic twined weave — punctuated with embossed diamond, diagonal, or zigzag patterns produced with special weaves — and with woven internal headbands to assure proper fit, the Haida spruce-root hat was truly a feat of fiber engineering. Although other Northwest Coast peoples made similar hats, the Haida product came to be highly valued by neighboring tribes as an item of trade. Such a hat produced today can easily command a price of $1000 or more for the weaver.

The most beautiful Haida hats were the prized possessions of the traditional upper caste of Haida society and often bore distinctive cylindrical ornaments — woven over cedar wood forms — atop their crowns that announced the number of potlatches that had been given by their wearers. Delores Churchill believes such a hat could even serve as a binding legal document, effectively committing its owner to carry out the wealth display feast that it had so boldly already recorded. She has also made the interesting observation that the bold horizontal black bands that decorate so many old Haida baskets — particularly the common, large cylindrical utilitarian ware — may have represented, like the ornaments on top of the Haida spruce-root ceremonial hat, the number of potlatches that had been given by the individual who owned the baskets. The ceremonial spruce root hat also provided an interesting exception to the general rule of basketry decoration: it was often painted with the same totemic animal figures, often in conventional reds and blacks, that distinguished Haida wood art and two-dimensional drawings. This colorful adornment was customarily painted, not surprisingly, by a Haida man.

The features that distinguish Haida basketry from the work of neighboring peoples are not always obvious today. In the first place, the Haida were active travelers and traders, and often intermarried with members of other cultural groups nearby. So basket-weaving techniques — in the form of objects of trade and the skills of new wives — inevitably diffused across ethnic boundaries. Add to this the sometimes questionable identification of Northwest Coast baskets by collectors lacking professional training (or perhaps unaware that a basket could surface in a place quite distant from its cultural origins), and any rigid system of classification of Haida basketry becomes even more elusive.

In an effort to resolve this uncertainty, Joan M. Jones, an anthropologist, undertook a computer analysis of nearly 3000 identified Northwest Coast baskets from museums all over North America. Despite the artificiality imposed by such a statistical study of somewhat arbitrary design elements, several diagnostic trends were identified in most Haida baskets. Among them were the frequent use of cylindrical basket shapes, braided rims, and relatively plain basket exteriors that reflected a preference for ornamentation created by the primary horizontal weave alone, rather than by secondary overlay of colored strands. Furthermore, Haida basketry during the past two centuries has been relatively conservative, catering less to the commercial whims of the European buyer than did the wares of aboriginal neighbors. In addition to simplicity of design, Haida baskets generally show a preference for robustness: "The Tlingit did finer basketry," states Delores Churchill. "Haida baskets were usually made to be *used*." This she attributes, in part, to the greater geographical isolation enjoyed by the offshore Haida.

According to Bill Holm, the respected curator of Northwest Coast art at the Thomas Burke Memorial Washington State Museum, Seattle, the clearest distinguishing feature of Haida basketry is the direction of the steplike shifts that appear wherever there is a change in a basket's weave or color. These steps contrast with those found in baskets made by neighboring peoples and are simply a consequence of the Haida custom of inverting the basket as it is being woven. Holm states that the occurrence of this "Haida jog," as he terms it, is consistent with the other characteristics of Haida ware that have been observed, such as the black bands, three-strand weaving on the rim, and cylindrical form.

Haida basketry and other native arts were changed forever with the arrival of the first European visitors to the Queen Charlotte Islands. The Spaniards reportedly reached the archipelago by 1774 and found a population of some 8000 Haida living a traditional way of life. Wave after wave of other Europeans followed, hungry for sea-otter pelts and salmon, "heathen" souls and valuable minerals, and, most recently, stands of virgin timber. These outside influences drastically affected, of course, far more than the arts of the Haida people: the foreign economy shattered the fragile surpluses that were the basis of the Haida potlatch ceremony, the mainstay of wealth and social status in Haida society, and an 1884 Canadian law even outlawed the potlatch for a time. Residential missionary schools often cruelly imposed new values on Haida children, isolating them from traditional family and social influences. Introduced diseases, such as smallpox, ravaged the Haida in an appallingly deadly series of epidemics. And, in their wake, alcoholism and other debilitating social ills emerged as problems whose effects linger still.

Also, pans of brass and tin, pewter basins, and textiles appeared in the homes of Haida basket weavers, displacing forever the tediously built utilitarian basket ware. And in exchange, Haida weavers turned to styles of Haida basketry that, serving as showpieces of the fiber artist's virtuosity, would command the highest export price. Although to a lesser extent than their Tlingit counterparts, Haida basket makers began to incorporate decorations that copied Christian symbols and European textile patterns, and occasionally to weave shapes — Communion goblets, cigar pouches, even tea kettles — that would please the foreign buyer and bring much-needed cash into the Haida home. The new trade baskets were fundamentally derived from the old, their shapes often elaborations of rounded traditional forms, their decorations — although supplemented with a new spectrum of synthetic colors and borrowed techniques of overlay — often rooted in the standard array of Haida twined weaves. But some traditional styles, such as the sturdy, undecorated spruce root and cedar bark ware that had dominated Haida basketry, approached extinction. By 1910, according to Erna Gunther, virtually no baskets were being constructed for use in Haida homes.

In recent years Haida basketry has enjoyed a certain renaissance, including a renewed interest in the full range of its earlier traditional forms. The recognition that older weavers are beginning to receive may encourage the recruitment of younger Haida to the art, reopening channels of communication through which certain Haida cultural values can be passed.

Selina Peratrovich, in her 90s, tells an old story of the first Haida basket, a story that stresses the qualities of generosity and loyalty to kin that are still vital in Haida culture. It tells of a time, long ago, when the Haida did not have enough food to eat, and of a stern Haida woman who scratched her daughter's face for eating more than her rationed share of food.

The young girl fled with her older sister to a beautiful place by a lake, where they encountered, in the form of a man, a Haida spirit-helper. Seeing their distress, he instructed them to "pull up spruce roots, split them, and weave them around your thumb" to make a basket. In it, they were told to place one of each kind of food that grew abundantly nearby.

Magically, the basket began to grow — along with its generous bounty of food — and the two girls returned to their village where they were greeted with great joy. The two girls went from house to house to share their precious basket of food with their neighbors, arriving last at the home of their mother. She too was given a gift of food, but died at once, it is said, of shame as she received it from the daughter she had previously punished so severely.

1982

Talaiden: A Sacred Law against Laughing at Animals

IN THE WORLD of the Chewong people — one of numerous Orang Asli, or "original peoples" who have long inhabited the now dwindling, biologically rich tropical rain-forest ecosystems of peninsular Malaysia — *all* species of animal are inherently deserving of profound human respect. The very existence of the Chewong, like so many other Malaysian Orang Asli peoples, is today imperiled by the modern, industrial non-native society that surrounds and threatens to engulf them. The beleaguered Chewong, their traditional land base withering rapidly under the pressures of logging and encroaching urban populations, numbered no more than 131 desperate souls in 1979. Nonetheless, the power of some of their most cherished traditional values regarding human responsibility for and participation in the natural world, to the extent we are aware of it, remains undiminished.

Nowhere is this more evident than in the ancient Chewong prohibition against ridiculing animals, embodied in a series of sacred laws known collectively in the Chewong language as *talaiden.* Stated simply, one of the major tenets of *talaiden,* concerned specifically with the notion of proper human attitudes toward other animal species, mandates that "no animal whatsoever may be teased or laughed at."[1]

If, for example, a group of young children in a Chewong village is found playfully taunting a captive bird or snake — or even if, for that matter, they are behaving too boisterously in the vicinity of animal flesh that is in the process of being prepared, cooked, or eaten — an adult member of the community is likely to step forward, scold them sternly, and insist that they stop their activity at once.

"*Talaiden!*" the scolder might hiss to the careless contingent of Chewong children. "The snake will gobble you up!"

The ravenous "snake," whose authority has been invoked here as a possible source of severe punishment for unthinking acts that offend the spirits of wild animals — captive or slain — is no ordinary reptile. Rather it is none other than Talo'den asal, the Original Snake, as every traditional Chewong child learns early in his or her life. Talo'den asal — the huge, coiled, primordial serpent that lives in the dim, watery depths of the underworld beneath the familiar hot, humid, and sun-drenched terrestrial realm of everyday forest life known to the Chewong simply as Earth Seven.

The Original Snake has a dual manifestation: it embodies, according to the Chewong, both a reptilian and a feminine principle. That is, the creature is best described as a supernatural woman in snake form, possessed of a strength and size so breathtaking that with no more than the idle twitch of her muscular body, she can unleash catastrophic, life-threatening, natural disasters. And it is this chaos that she metes out as punishment to those human beings who, as she senses in her silent omniscience, have through one act or another violated the sanctity of some member of the all-encompassing diversity of animal species under her care.

A wrathful Talo'den asal might, for instance, retaliate by releasing volleys of thunder, lightning, and wind in a fierce tropical rain storm. By so doing, she is able to make towering, shallow-rooted trees near the offending Chewong community tremble and bow until they crash to the rain-drenched forest floor with the impact of an explosion. Or she might cause waters from reservoirs in the underworld to begin to seep upward, where they can serve as a threat to the Chewong of an imminent world flood, should such human infractions against the well-being of the tropical rain-forest fauna persist.

But it must be emphasized that despite her awesome destructive powers, Talo'den asal, the Original Snake, is not viewed by traditional Chewong, as some might suspect, as either an embodiment of evil or a malevolent force of nature. Rather, they seem to look upon her as a reflection of that same wondrous mix of complexity and mystery that characterizes the totality of the natural world itself.

That is, she is — like Tanko, a complementary Chewong figure, who serves as her masculine counterpart and spouse and dwells high in the cloud-flecked skies over Earth Seven — steeped in the same powers and parodoxes that the Chewong encounter in nature. In fact, at times, Tanko's behavior, like hers, suggests an unapologetic mix of comedy and tragedy, hilarity and grief. For example, he is known in myth to have occasionally displayed his mirth at the endless personal foibles and predicaments of mortal human beings acting out their petty lives below in the theater of Earth Seven by issuing, along with his punitive lightning bolts, raucous thunderclaps of heartfelt laughter.

Thus, both Talo'den asal and Tanko can be seen, in the words of Signe Howell, the only Western anthropologist to have lived long and intimately with the Chewong in recent years, at least in one obvious aspect of their multi-layered meanings, as "abstract symbols of nature" par excellence — nature, that is, in all its divine and unfathomable richness.

Openly laughing at or teasing an animal may represent the most blatant kind of human violation of the Chewongs' *talaiden* rules protecting other creatures. But there are also other, subtler kinds of transgressions against animals. Among them are ones that one can easily imagine traditional Chewong (if they enjoyed the luxury of frequent foreign travel) would almost certainly spot instantly in Western societies' ridiculous, often demeaning images of animals for human entertainment. They might easily see a worrisome breach of the laws of *talaiden* in Hollywood's long legacy of buffoonish cartoon creatures — ranging from a perpetually adolescent, bipedal song-and-dance mouse named Mickey to gangs of masked, mutant, karate-kicking Ninja turtles born of subterranean toxic oozes. They might justifiably invoke *taleiden* rules upon seeing beleaguered Ringling Brothers circus bears astride bicycles or wide-eyed monkeys dressed up in clownish human attire; stylishly coiffed pet poodles; or Siamese cats draped in sparkling rhinestone collars or destined to be buried in posh suburban cemeteries. To Chewong sensibilities, each of these situations would likely represent a genuine case, however unconscious, of a direct human affront to the inherent dignity of a fellow animal species.

Another aspect of the ancient Chewong notion of *talaiden* is this: a human being diminishes another species by treating it as if it were a mere plaything, a toy. One Chewong legend, recorded by Signe Howell in her insightful studies of the Eastern Chewong of peninsular Malaysia, reveals the precise nature and consequences of such a transgression.

A Chewong man and his fiancée happen to capture a live squirrel while they are walking in the tropical rain forest. Perhaps because they are young and still childless, they unthinkingly decide to bring the captive animal home to their village, where they plan to keep it as a pet.

The young couple momentarily forget their fundamental obligation to honor any animal they might encounter in the forest. They eagerly place the squirrel, though it is all but paralyzed with fear, in a string hammock originally woven to cradle a human infant, which hangs expectantly from the ceiling of their thatched hut. Then — treating the little creature almost as if it were the human baby of their shared dreams — they proceed to playfully rock it back and forth, murmuring all the while the same soothing Chewong words they intend one day to sing for their own offspring.

The deed is already done. Simply by *pretending* that their helpless captive is human, they have committed — as they are about to discover — a grievous offense against the squirrel and all its multitudinous kind. For they have in effect denied the animal its fundamental right to its own natural identity and place in the cosmos. And, thus, they have, however gently, broken one of the rules of *talaiden.*

Talo'den asal is aware of their flagrant breach of this taboo the very instant it occurs. She awakens with a start from her dark subterranean slumber and flies into a rage over the plight of the squirrel above in Earth Seven. Twisting her long, multi-hued body into a coil of fury in response to this wrong, she suddenly unleashes a terrible storm, as if it were a spear, into the Chewong village above her. Howling winds, torrential rains, and a rising tide of groundwater all converge on the young couple's hut, sweeping it from the face of the Earth as if it were no more than a speck of dust.

In a final act of horror, the great serpent rears up from the cavernous underworld, grips the two offending Chewong viciously in her great maw, and gobbles them both up in a single predatory gulp. But Talo'den asal does not simply devour them and leisurely digest their flesh and bones, as she might ordinary prey. Rather, she *obliterates* their very existence for all time. As she swallows them, the supernatural reptile deliberately snuffs out the fragile flame of vitality that flickers in, and animates, not only each human being but, equally, *all* forms of life on Earth Seven. And by righteously extinguishing the precious *ruwai*, or "soul," of each of the young Chewong, she assures that these will not continue to glow on in realms beyond Earth Seven long after the couple's deaths — as they would have in the absence of so grave an offense against nature.

What is the message of this modest Chewong tale? According to anthropologist Howell, its meaning may well be at once exceedingly simple and ecologically profound. It is this: "Different species of animals [including *Homo sapiens*] must be kept separate according to their characteristics." In other words, "each has its own role to play, and proper behavior on the part of humans *vis-à-vis* them must be observed."

There is also an intriguing corollary to this Chewong maxim — one with, it seems, obvious potential application to certain laboratory techniques in modern biotechnology and genetic engineering involving the casual swapping of genes from one species to evolutionarily distant others. Contemporary Chewong believe, putting their faith in hundreds of generations of accumulated experience in tropical rain-forest ecosystems, that *any* human attempt to foster unnatural matings between distinct species of animals (the mere idea today, for example, of somehow "marrying" creatures as patently distinct as a cat and a dog) is a violation of timeless laws protecting animals. That is, in the traditional Chewong cosmos, *all* species — not just *Homo sapiens* — are imbued with a uniqueness, an identity, and an inviolability that render them equally sacred as members of the natural world. And as a consequence, human beings have a solemn duty to recognize that, in Howell's poignant phrase, the "boundaries between classes of beings must be upheld."

1990

1. Signe Howell. *Society and Cosmos: Chewong of Peninsular Malaysia*, Chicago: University of Chicago Press, 1989.

Part IV

ON THE NATURE OF HUMAN DIFFERENCES

Introduction

IT HAS BECOME quite fashionable these days to speak earnestly over cocktails or cappuccino of the inherent value of maintaining *biological diversity* — that is, in preserving as much as possible of the rich variety of different plant and animal species that evolution has bequeathed our planet, especially within the tropical rain forests of Latin America, Africa, and Asia, as well as other ecologically imperiled regions of the world. And that, of course, is to be commended. But at the same time there often seems to be disturbingly little discussion in some of these same circles of the simultaneous loss of *human* diversity — that is, the diminishing reservoir of both cultural and biological variety expressed within *Homo sapiens*. And that, I think, is cause for concern.

I know of few more striking examples of one people's systematic sense of biological and cultural superiority over another group than in the historic confrontation between aboriginal peoples of the New World and European colonists of the Old — beginning with the arrival in Americas of Spanish conquistadores almost exactly 500 years ago.

In 1988, with the aid of a Canada Council grant for non-fiction writing, I set out to explore aspects of this topic, focusing upon the history of post-contact oppression endured by the Maya Indians of Guatemala and southern Mexico not only at the hands of the first Spanish invaders but also under successive waves of European and North American economic, military, and political exploitation of the Maya that continue as I write.

The prevailing pattern of post-contact Maya history is shockingly clear to anyone who cares to probe beneath the "official" story. It is this: over the past five centuries Maya peoples of Central America have been the victims of a continuous policy, in many guises, of racist political violence, religious persecution, and cultural intolerance on a scale so vast that it can legitimately be called physical and cultural *genocide*. The Maya's ongoing story of suffering and resistance at the hands of foreigners is in a very real sense, as a Maya friend living in Canada recently reminded me, one that deserves comparison to the Holocaust.

That tale begins perhaps in 1502 (a full decade after Christopher Columbus's first New World sojourn) with the first recorded contact between the Maya and Europeans off the coast of the Yucatan peninsula — an encounter that was apparently peaceful and full of promise. But by 1524, following Cortes's stunning victories over the Aztecs in Mexico, a small army of Spanish conquistadores, under the command of ruthless red-bearded captain Pedro de Alvarado and equipped with state-of-the-art Old World weaponry, was busy cutting a bloody path through the patchwork of post-Classic Maya city-states located in what we today know as Guatemala.

In the wake of Alvarado's murderous assault would come a series of additional mortal blows to the Maya that would eventually all but bring this ancient, literate, and extraordinarily sophisticated New World civilization to its knees. There were raging epidemics of smallpox and other introduced infectious diseases, brutal military occupations, and a blatant Indian slave trade, in which Maya men, women, and children were routinely scarred with state-sanctioned branding irons as proof of "ownership" and sold like cattle. And there was the predictable creation, by the Spanish Crown and the Catholic Church working in tandem, of a racist and ethnocentric colonial system founded on terror, economic exploitation, forced Indian labor, absolute political and social control, and relentless religious persecution — the remnants of which linger, unmistakably, despite independence from Spain in 1821, in twentieth-century Guatemala.

The Maya's story would also point out the complicity over the past century of other powerful foreign interests besides Spain — among them German, British, and, most recently and prominently, American — in consciously contributing to the oppression of

Maya people. It would document the pivotal role of the American CIA covert operation SUCCESS, with the help of U.S. corporations in the region, in orchestrating the infamous 1954 coup that destroyed Guatemala's brief decade-long flirtation with democracy and meaningful rural land reform. It would, justifiably, hold the U.S. government accountable for the crucial part it has played in propping up and manipulating a succession of right-wing military dictatorships as brutal as any that Latin America has witnessed during this century. Not to mention United States complicity, along with that of Israel, in providing equipment, training, and expert advice to the corrupt Guatemalan army. That military institution remains, in its Faustian alliance with the nation's wealthy, predominantly European and Ladino (mixed blood) elite, to this day, despite smoke-and-mirror claims of the emergence of fledgling "democracies," the mainstay of a suffocating, racist status quo for the Maya.

The outside world continues, for the most part, to respond with unpardonable silence to the Maya's tragic, ongoing tale of woe. Part of the reason for this is the fact that reliable accounts of the plight of Maya peoples in Guatemala have been for a variety of reasons virtually absent from mainstream North American media, despite the indisputable role that U.S. foreign policy and business interests have played in prolonging the agony of the Maya through much of this century. And, to compound the tragedy, this grim, five-century-long genocidal chapter of Maya history has, to date, yet to be fully recounted to the world in the cries of a truly Mayan voice.

In this section, I have included one brief excerpt, titled "*Matanza:* The Massacre of Maya Indians at Finca San Francisco, Guatemala," from a book in progress. The book represents my own attempt to try to trace the roots of the modern apartheid-like political and economic oppression of Guatemala's Maya — who, like their more visibly distinct black counterparts in South Africa constitute an indisputable majority in that country — in the simmering racist persecution that has dominated that region since the sixteenth-century European invasion.

It is a meticulously documented reconstruction of an actual military assault upon an entire unarmed Maya community by government counterinsurgency forces during the early 1980s, at a time when the seemingly endless cycle of violent repression in

Guatemala happened to have reached one of its many peaks. It is an attempt to compel compassionate non-Maya readers to begin to respond to the nightmarish dimensions of Maya suffering with genuine empathy. It is also meant to stand as a case study of long-standing Maya claims of genocidal treatment. I refer not just to specific acts of genocide that have taken place over the past several decades — in the form, for example, of wholesale torture and terror of Maya civilians, political assassinations and "disappearances," forced relocations of "internal" refugees and massive exoduses of "external" refugees — at the hands of a succession of military-dominated regimes shaped, in large measure, by the U.S.'s paranoid post–Second World War obsession with destroying whatever it chose to define as "Communism." Rather I refer to genocide, physical and cultural, as *the* central fact of post-contact Maya history — a fact that few Maya will forget during the public 500-year "celebration" by non-native peoples, scheduled for 1992, of Columbus's historic "discovery" of the "savage" New World.

I also include in this section on "The Nature of Human Differences" an article that focuses particularly on the relationship between science, rather than society as a whole, and racism. "Rushton on Race" is an unpublished essay on an unfinished Canadian controversy that broke into the news in early 1989 and has already reverberated far beyond Canadian borders. The topic of the shrill public debate, initiated by a previously obscure Canadian psychologist named Philippe Rushton, is hardly new to historians of twentieth-century science: it is another hypothesis of possible biological bases for such value-laden human traits as intelligence, sexual behavior, and health.

The Rushton piece was originally commissioned by *Equinox* magazine. Earlier versions of the manuscript have been reviewed — all or in part — by E.O. Wilson, Stephen Jay Gould, David Suzuki, and Rushton himself, among others. But I of course assume final responsibility for its contents and conclusions. For reasons that I still have not fully fathomed, the editors of *Equinox*, as well as, in turn, *Toronto Life* and *Saturday Night*, chose not to publish it. I, in contrast, still consider it rather important, so I offer this unexpurgated, perhaps at times slightly uncomfortable, tale of what I have come to view as academic racism's latest incarnation — here, in our midst, in the gentle heart of Canada.

Matanza: The Massacre of Maya Indians at Finca San Francisco, Guatemala

Genocide: "Any act perpetrated with the intention of destroying, totally or in part, a national, ethnic, racial, or religious group . . ."
U.N. Convention for the Prevention and Punishment of the Crime of Genocide, December 9, 1948

During a span of six centuries, from about A.D. 250 to 900, the Maya, particularly those of the Central Area, reached intellectual and artistic heights which no others in the New World, and few in the Old, could match at that time. . . . [Yet] the Maya are hardly a vanished people, for they [represent] the largest single block of American Indians north of Peru. Most of them resisted with remarkable tenacity the encroachments of Spanish American civilization, although in the past decade these have taken an increasingly violent and repressive form.
Michael D. Coe, in The Maya *(4th edition), 1987*

The official news in Guatemala never divulges what is really happening. In fact, hundreds of indigenous people have been massacred in all of the Indian provinces. Many Indian villages, especially in the highlands, have been wiped out. Children, women, elders, and even babies have been killed. These massacres, killings, and kidnappings were committed by the army and paramilitary groups paid by the landowners and the government itself . . . More than 250,000 people, mostly

> *Mayan Indians, have fled to the southern part of Mexico and are living in refugee camps. Inside the country, more than one million people have been displaced and are, in fact, "internal refugees." Thousands of children have been orphaned and as many women widowed.*
>
> *Mayan Indian Support Group,*
> *an organization of Maya refugees,*
> *Vancouver 1986*

Prologue

In all of Latin America, only four countries — Guatemala, Bolivia, Ecuador, and Peru — can lay claim to aboriginal populations that represent clear political majorities. In the case of Guatemala, that claim has not surprisingly been disputed, largely because of the vagaries of demographically defining precisely who is or is not a Maya Indian. In 1984, the Guatemalan government (one in a long line of military-backed regimes that have succeeded one another since the United States orchestrated the military overthrow of an elected democratic Guatemalan government in 1954) maintained that native Indians constituted less than 40 per cent of the country's residents. Today, more reliable observers maintain there is absolutely no doubt that the Maya constitute a substantial majority — at least between 55 and 65 percent — of a Guatemalan population that in 1990 was reported to be more than nine million. The Maya themselves — most of them members of the Quiche, Cackchiqueles, Mam, and Kekchi Maya nations — insist that their true numbers are far greater — perhaps as high as 75 or 80 percent of the total population.

One thing is certain. In the nearly five centuries since the first trickle of Spaniards arrived in Mexico's Yucatán Peninsula, the Maya have endured relentless poverty, misery, racism, and cultural loss — with virtually no political voice in the affairs of their country and little moral support from the outside world. During the past decade, in a land increasingly torn by the polarization between the Guatemalan military and a leftist guerrilla insurgency dominated and controlled by a Ladino leadership, this oppression has dramatically escalated. Beginning in May 1978, when government troops massacred 100 Kekchi Indians and wounded 300 more during a peaceful protest in the town of Panzos, the Maya and their way of life have come under increasingly violent attack. In an effort to terrorize the rural Maya population and neutralize them politically, a succession of military regimes have subjected them to relentless counterinsurgency programs that have often borne an uncanny resemblance to American strategies used in Viet Nam during the 1960s. They have been described as perhaps the most cruelly efficient in all of Latin America, if not the world.

According to Amnesty International, at least 100,000 people, most of them Maya, have been murdered or "disappeared" in Guatemala since 1954 — more than 30,000 since 1981. (It has been estimated that almost 40 percent of all "disappearances" throughout Latin America since 1966 have occurred in Guatemala.) In addition, more than 400 Indian communities have been deliberately destroyed by government "pacification" programs. By 1984, more than one million Guatemalans — again, mostly Maya — had been uprooted from their lands and turned into refugees in their own country. Another 250,000 had been forced to flee Guatemala entirely. Most arrived in refugee camps in Mexico, El Salvador, Belize, or Honduras only after perilous overland treks that left them ill, exhausted, and fearing for their lives. Some managed to make their way north — legally and illegally — to cities in the United States and Canada.

Burgeoning Maya Indian communities in Miami, Los Angeles, San Francisco, Seattle, Vancouver, Toronto, and other North American cities are living testimony to the magnitude of that explosive period of forced migration. In fact, this historically unprecedented flight of the Maya from their traditional lands over the past decade stands as perhaps the most visible symptom to outsiders that there exists in Guatemala an indisputably racist, anti-Indian political system. One that — were the contrast in skin pigmentation between politically powerless Maya Indian and privileged European descendants or mixed-blooded mestizo as striking as that between South African black and South African Caucasian — would readily be recognized by the world for what it is: a distinctively Latin American version of apartheid.

SATURDAY, JULY 17, 1982, is only one of countless days of infamy and grief etched in the collective memory of Guatemala's Maya. On that day a terrible massacre, or *matanza*, took place in an obscure Maya community called Finca San Francisco. *Matanza* is the Spanish word for the indiscriminate killing of human beings; it is also the word for slaughtering or butchering domestic animals. *Finca* means "estate" or "plantation"; San Francisco is simply the name of the estate. The massacre at Finca San Francisco occurred only weeks after General José Efrain Rios Montt unilaterally declared himself president of Guatemala and formally declared a national state of siege, following a right-wing military coup staged, in the wake of an abortive election, on March 23, 1982. This was followed by an exceptionally murderous time that has been aptly described as "Guatemala's bloodiest period of rural repression in the twentieth century." Rios Montt's actions represented a desperate attempt not only to crush what had become an increasingly bold guerrilla movement in the highlands but also to silence potential future resistance to his regime on the part of Guatemala's impoverished Maya majority.

The massacre at Finca San Francisco is, in a sense, a microcosm of the sheer brutality of that period. During that nightmarish 24-hour period more than 300 innocent Maya men, women, and children were brutally murdered. They were executed en masse — not because they represented any sort of direct threat to Guatemala's military regime or vital link to the armed guerrilla movement in the region, but, in a very real sense, simply because they were Indian. That is, as Indian people, they were — despite constituting a legitimate, if voiceless, majority — viewed by the powerful Guatemalan elite as categorically subversive and expendable, and valued, as the early post-contact history of Maya-European relations clearly documents, as essentially less than human.

More important, by selectively eliminating entire Maya communities such as this one in a single bold stroke the Guatemalan army apparently hoped both to demonstrate its military might and to frighten rural populations into continued submission. For as word of wholesale Indian massacres spread among impoverished rural Maya laborers and subsistence farmers — so the cold logic of international anti-Communist counterinsurgency goes — Maya resistance, in any form, against their oppressive government would be extinguished before it had the chance to burst into full revolutionary flame.

The bloody massacre of an entire unarmed Indian community

at Finca San Francisco in 1982 does not stand as a singular act of racist violence against the Maya in Guatemala. But it is unique in one important way. Because of the extraordinary efforts of Ricardo Falla, a respected Guatemalan Jesuit priest with a doctorate in anthropology, to document for posterity every excruciating detail of this incident, the story of the final day in the life of this Maya community has been recorded for the whole world to see.

The narrative that follows is based almost entirely on Ricardo Falla's highly regarded report on the massacre,[1] and the direct quotations that appear come from the English translation of testimonies of surviving eyewitnesses of what today stands as perhaps the most visible — yet hardly the most inhumane — act of "Mayacide" in the troubled history of Guatemala. I salute his efforts and acknowledge my debt as a writer to Falla and especially to his courageous Maya informants.

These events took place, as I have mentioned, on July 17, 1982, in a small farming village located in the remote western highlands of Guatemala, not far from the Mexican border. The chronicle of that 24-hour period reveals — if one looks closely and permits oneself to feel deeply — elements of the nightmarish political oppression, stunning disparity of power, and vicious racism that have in one guise or another characterized the nearly five-century-long domination by colonial Europeans and their descendants of Guatemala's indigenous Maya peoples.

The purpose of reliving that day in Finca San Francisco is not simply to shock readers. Rather, it is to provide them with a faithful glimpse of events in very recent Guatemalan history, to the extent it is possible, through the eyes and ears of Maya who witnessed them. Unless non-Maya people are willing to look directly into the face of such monumental cruelty and make an effort, however uncomfortable, to empathize with fellow human beings who have been its victims, there is little hope they will even begin to comprehend the grim day-to-day *realidad*, or reality, that arises simply from being born Maya in a Guatemalan world.

It is also aimed at arousing in North American readers a measure of indignation, however long overdue, at the by now documented role of the United States in promoting — overtly and covertly, for its own political and economic interests — an atmosphere in Guatemala in which political violence and abuses of human rights have long flourished. [2]Beginning early in this century, the United Fruit Company — the very epitome of twentieth-century American corporate colonialism in Latin America — openly pillaged the Guatemalan economy for decades. And by 1954, shortly after a portion of the company's huge land holdings had been expropriated, the CIA boldly engineered a coup, overthrowing a duly elected, if irritatingly nationalistic, Guatemalan democracy, and ushering in the blood-stained modern epoch of brutal military control. Even today, recent reports suggest that the U.S. investment in Guatemala is higher than in any other country in Central America, including Costa Rica. In 1985, for example, a total of 208 U.S. corporations were making direct investments in Guatemala of $325 million (U.S.) annually.

There is more. Since the 1960s, Guatemala has received over U.S. $66 million in military aid. And at times, thousands of Guatemalan army officers — among them future dictators — were routinely trained at American military institutions. Even during periods when military assistance to Guatemala was officially suspended, the U.S. government, along with Israel and other countries, managed to quietly inject millions of dollars of so-called "non-lethal" aid and expertise into the police and security systems of corrupt Guatemalan regimes in a country that Robert Carmack, a noted American authority on the Maya, has described as possessing "one of the worst human rights records in the Americas."

As you read what follows, consider what your own feelings might have been if the identical events had taken place in your own country, in your own community, to your own loved ones. To facilitate this leap of the imagination, I have included at the end of my narrative a list of the names and ages of the known dead at Finca San Francisco. It is also my way of giving them a measure of the respect they were in life denied — as well as the final word in their own story.

The Massacre

Approximately 11:00 a.m.: The first ominous sign of impending violence occurred that morning, according to eyewitnesses, when approximately 600 Guatemalan army foot soldiers entered the community, which was situated on the 180-hectare agricultural estate owned by an army officer by the name of Colonel Victor

Manuel Bolaños. The soldiers, dressed in the distinctively patterned camouflage uniforms that had earned them the somewhat facetious nickname *pintos*, or "speckled ones," among many rural Guatemalans, arrived from the southwest, along the dirt road that led to the neighboring village of Yalambojoch. At about the same time a Guatemalan army helicopter, painted white and laden with food and supplies for the soldiers, loomed suddenly overhead and circled low over San Francisco before settling noisily on the village soccer field.

The villagers took notice of the army's arrival, to be sure, but they did not appear to be unduly surprised by the sight of the approaching soldiers. After all, cattle belonging to the *finca* had recently been subjected to a series of raids by roving bands of hungry leftist revolutionaries — members of the Guerrilla Army of the Poor, which was one of the principal factions of the armed opposition to the Guatemalan military government. In addition the main house on the estate had been burned down by guerrilla forces the previous December. More comforting, there had been no word recently from neighboring villages of any violent reprisals by government troops against civilians suspected of sympathizing with the guerrillas. As a result the residents of Finca San Francisco had grown somewhat accustomed to the movement of government troops through the area. And for its part the army had at times openly courted the favor of Indian populations in the region, including the one at San Francisco — though not always with complete success.

Still, among the townspeople were some who sensed almost instantly that this military maneuver might be something more than just another routine visit by the Guatemalan army. For one thing, as the troops drew closer to town they noticed that faces of some soldiers were contorted in anger. "We were watching them," observed one eyewitness later. "Their faces were like those of crazed men. It was powerful!" For another thing, they could see that several soldiers were leading a lone prisoner by a rope. He was a man dressed, like his captors, in the familiar camouflage-style khaki of an army *pinto*. But he was, recalled a villager, "tied with a soldier's belt, like a dog."

It soon became apparent that the prisoner, presumably a captured guerrilla, was being forced by the soldiers to identify possible guerrilla sympathizers in Finca San Francisco. At one point, when the prisoner pointed to a particular peasant farmer among the assembled, the soldiers quickly dragged the unfortunate man forward, bound his hands, then proceeded to cut a bloody slash across his face. This sudden display of force against one of their own seemed to stun the villagers, causing one to turn to his neighbor and whisper softly in Chajul, the local Maya dialect, "They are going to finish him off. Now we are screwed."

Several soldiers ordered two boys from the village to fetch a pair of bulls for the troops to eat, specifying that the animals should be taken from impoverished Indian peasants in San Francisco, not from the army officer who owned the ranch. "Now, we want you to bring two head of cattle for us to eat," they said. "We've come here for you, you see; we are going to have a *fiesta* here with you right now." By the threatening tone of the soldiers' voices and their chilling reference to a joint "celebration," despite their recent display of hostility, some villagers immediately assumed that things could only get worse. "We started realizing then that they were going to kill us," remembered one. Despite the injustice and innuendo in the demand, the villagers delivered the cattle to the soldiers without delay.

Meanwhile, the colonel in charge of the military operation issued a command that the townspeople should assemble immediately for a public meeting in the center of town. Soldiers were quickly dispatched to pass this message on to any residents who were not within earshot of the announcement. "There is going to be a meeting," the villagers were told. "The colonel wants to have a discussion with you; go and hear him, and make sure you listen carefully."

Before long most of the town's population — more than 300 Maya men, women, and children — stood anxiously in the plaza of their tiny village to await word from the army of its intentions in San Francisco that day. Despite the unnerving events that had already taken place that morning, some villagers remained hopeful that the violence would not escalate any further. Recalled one: "Since we did not know what wrong we had done, we did not think that they had come to kill us. The first time [the army came], they had told us not to fear them, so by then we had begun to trust the soldiers."

At the public meeting, events seemed to begin to take on a life of their own. First, all the adult men from San Francisco were ordered inside the courthouse without explanation, and the doors were locked. The women and children were then herded inside the church some 20 meters across an open plaza from the men. "They did not say what they were going to do with us," recalled one survivor later. "Who could have known what was on their minds? We were beginning to realize, but how could we have escaped? They had us surrounded."

In a bizarre turn of events, as soon as the residents of San Francisco had been imprisoned, the troops took a break in their sinister work to indulge in a meal. Soldiers proceeded to slaughter one of the bulls the villagers had been forced to give them earlier, bloodying the patio of the village school. They skinned the animal, butchered it, and roasted it over a fire. Finally, they feasted on the freshly killed beef, wrapping it in tortillas made by local women, which they had pilfered from homes in the village. It was as if the only reason for their sudden visit to San Francisco had been for an impromptu picnic.

After eating their fill, a group of between 70 and 80 soldiers departed from the main body of soldiers and began to ransack the modest huts belonging to the poor farmers of San Francisco. As they searched through the sparse personal belongings in each home, they stole anything that struck their fancy. In the process, they also stole an estimated 10,000 Guatemalan *quetzales* (at the time, one *quetzal* was worth approximately one U.S. dollar) from the treasury of the local community cooperative and an additional 10,000 *quetzales* from precious individual cash savings, along with an assortment of legal documents and personal papers.

"They robbed us of everything we had," one villager would recall bitterly. "They stole my wristwatch, 20 *quetzales* from my wallet, and even my nail clipper. They left us with nothing. Even though we are *campesinos*, we manage to get the things that we need in this world. We always have a little money in the house. But they took that, as well as radios, cassette recorders, watches, and clothes."

Midday: by noon, all the people of San Francisco — men, women, and children — except for a few fortunate farmers who happened to be on errands outside town or laboring in distant fields, were held under armed guard either inside the church or the courthouse.

As sounds of the rampaging soldiers reached the ears of the captive villagers, anxiety mounted. Babies cried. Young, wide-eyed Indian children gazed up silently at their terrified mothers or cried softly to themselves. A few elderly men and women stared vacantly at walls, nervously fingering their canes. Some people prayed. "Now it is time to pray to God," whispered one Catholic lay worker. "We have to remember our commitment to God. If we have to suffer this punishment, there is nothing else to do." Others seemed to be already slipping into a state of shock, as if in anticipation of the violence to come. Recalls one witness: "We were all inside just waiting. Maybe we were not thinking anything, just 'now we are finished.'"

Approximately 1:00 p.m.: The sharp crackle of gunshots ripped through the uneasy calm that had hung over Finca San Francisco during the brief few hours since the arrival of the troops. More shots followed as Guatemalan army soldiers continued to fire their automatic weapons directly into a crowd of defenseless Maya women held inside the church. The burst of gunshots were followed by a noisy commotion and the shrill screams of women and children.

Across the plaza in the town hall the captive men could only catch fleeting glimpses of ensuing events through a hole in a shuttered courthouse window. But they could hear everything. After the soldiers ceased firing, they began to round up the remaining women into small groups of about 20. They then methodically led each group out of the dim interior of the church into the harsh afternoon light, forcibly separating mothers from their wailing offspring.

Each group of women, in turn, was herded at gunpoint into an empty house nearby. Then a number of soldiers entered the houses and raped the women. Following this, the women were ruthlessly slaughtered — by a hail of bullets or the explosion of an accurately thrown hand grenade, or sometimes with slashing machete blades. After massacring a group of women, the soldiers set fire to each house, gradually transforming rows of rude, dirt-floored huts piled high with the dead and dying women of San Francisco into a series of blazing funeral pyres.

To the growing horror of the men in the courthouse, the bloodied soldiers then returned to the church to retrieve the children, ranging in age from infants and toddlers to young teenagers. Among the imprisoned men were a few who were able to witness the mind-numbing atrocities against the Maya children that were to follow. Recalled one eyewitness simply: "They beat the hell out of them."

First, the men watched soldiers carrying babies out of the church and tenderly cradling them in their arms, only to suddenly stab them with their knives or nonchalantly bludgeon them to death. Then they watched soldiers catch toddlers, grab them by their legs and smash them like rag dolls against poles or trees, crushing their skulls. Some soldiers even disemboweled their helpless victims, slicing open their bellies with knives, extracting long, pale, glistening ropes of stomach and intestine, and casually flinging the organs into the dirt, as if they were nothing more than the offal of a young calf freshly butchered for market. In one such incident, according to the testimony of an eyewitness, "They killed two boys with knives. Those *cabrones* [bastards] cut one child's stomach like this," he said, making a slashing gesture across his abdomen," and tossed it away. By two in the afternoon it was over.

When they had finally completed their murderous task, the soldiers stacked the limp bodies of the children of San Francisco in bloody heaps inside houses. A few of the youngsters — though mortally wounded — continued to emit haunting cries and groans. But inside the houses, soldiers quickly silenced any of them that showed signs of life. "They cut out the intestines of the poor little children," confirmed another survivor. "Even after that, the children kept crying. The soldiers piled them up inside a house and finished killing them."

For the next hour there was a temporary lull in the carnage at Finca San Francisco. Inside the courthouse, the men of the village, many of them rendered childless and widowers in recent minutes, agonized over the unspeakable killings they had just witnessed. Outside, soldiers rested or milled about aimlessly. Again, almost unbelievably, some paused to busy themselves in the preparation of a meal.

Approximately 3:00 p.m.: Following this brief respite, the simmering violence of the Guatemalan army's rampage in San Francisco erupted once again. First, the soldiers entered the courthouse and dragged several elderly men of the village — the most revered members of Maya society — out of the building. These they slew roughly, as if slaughtering sheep, by slitting their throats with a rusty machete. Some of the soldiers laughed aloud as they carried out their gruesome work. When they had finished, armed soldiers began to lead small groups of males, each made up of 10 men of working age, their hands bound, out of the courthouse and onto an adjacent patio. Here, the captives were pushed to the ground, then executed by a gunshot at close range, with at least four bullets fired into the body of each man.

"The soldiers had begun. And it went on and on," one survivor observed. "I have to tell you what I saw; it does not matter any more. Everything was death already. The people were dying, dying. I was trapped like a fish in a net."

At one point, a soldier stooped over a man who had already been killed during the day, brandished a blade, and proceeded to cut open the man's body and triumphantly remove his heart. "The poor man was already dead, then the soldier attacked him again," recalled a witness. "He opened him up and took out his heart. I saw it, but I sat down. I could not look any more. I do not know if they ate it or took it with them."

Back inside the courthouse, the government soldiers had worked themselves into a killing frenzy. Raging with anger, they screamed at the remaining men huddled inside the building to step forward, one at a time, to be executed. And methodically, with a cold fury, they shot each of them, sometimes in full view of their captive neighbors. No one was spared, not even those who pleaded for mercy on the basis of local privilege or rank; the soldiers killed the town's officials, including the sheriff, the auxiliary mayor, and several local policemen. The manager of the *finca* was momentarily spared and tied up, only to be killed several days later.

At one point, a number of young Indian boys clung together in one corner of the courthouse, sobbing that they did not want to die. An older villager tried to calm their fears by suggesting that because of the painful suffering they had already been forced to endure and the execution that still awaited them, all of them would be forgiven for any misdeeds at the instant of death. "It will not last much longer, boys," the witness recalled telling them, "and now that we have suffered this punishment, we will go to the cemetery already pardoned." The terrified boys responded to these kind words by collapsing to the courthouse floor and huddling together for comfort. And the executions continued.

Following this series of systematic murders, the soldiers dragged the growing inventory of fresh corpses from the courthouse area and heaved them like sacks of flour inside the church.

Next, with perhaps 20 men and adolescent boys still trapped inside the courthouse, soldiers set fire to the building, apparently intending to incinerate the last citizens of San Francisco. Driven back by the searing heat of the burning building, the government troops gathered along one side of the structure to witness their final act of destruction.

Approximately 5:30 p.m.: Miraculously one of the remaining villagers managed to pry open a courthouse window facing away from the assembled soldiers. In quick succession, seven men squeezed through the opening and tried to sprint to safety. When the soldiers spotted the escapees, they leveled their rifles and instantly killed three of them. The remaining four men managed to elude their pursuers, fading quickly into the familiar terrain surrounding San Francisco. Enraged, the soldiers lobbed hand grenades into the burning courthouse in an attempt to finish off the villagers. They succeeded in killing all but two men.

Later, as dusk began to settle on the village and soldiers were stacking corpses inside the courthouse, they did not notice that two blood-soaked villagers were only feigning death as they were tossed roughly on top of the heap of broken and lifeless bodies.

Approximately 7:00 p.m.: With the village veiled in darkness, the two impostors attempted their desperate escape. As soldiers lounged outside, singing boisterously, playing music on stolen tape recorders, and counting their booty, the two slipped quietly out through the same window that their companions had used an hour and an half earlier. One inadvertently revealed his presence by making scuffing sounds with his boots. Almost instantly, he was sprayed with bullets. The other, who had removed his boots, crept unnoticed into the thick brush at the edge of the village. As he made his escape, he prayed silently to his dead family, friends, and neighbors, saying: "*Compañeros, compañeros*, let me go free to the fields. Give me luck. You are already free. Let me go! I am going to the fields."

The next day, hungry, weary, and in state of emotional shock, the survivors straggled across the Mexican border into the rural state of Chiapas, Mexico. One of the men, bearing a mortal wound as a result of the ordeal, died later in a Mexican hospital. About 20 other residents of San Francisco survived the destruction of their community simply because they were out of town for the day. These too eventually fled, along with thousands of refugees from neighboring villages, to the relative safety of Mexico. "That is how we saved ourselves. Everything was lost," explained one of the survivors later. "Everything was destroyed by the army. All our neighbors and our families were killed."

"We feel so much pain," said another. "The way they killed our people! How could the government order the soldiers to kill women and children who have committed no crimes? If they were looking for subversives, tell me, how many weapons did they find on the children? How many guns did the women turn in? That is why we are angry. How can this happen? The government is crazy. They are sending their army to kill people who have done no wrong. I do not understand."

An American priest living in the area was later approached by a man from a neighboring village who said that as soon as he spotted smoke billowing up on the horizon above Finca San Francisco he was certain that the Guatemalan army was burning the village to the ground. That evening the same man ventured to the ruins to determine what had taken place. All he saw, according to the American priest, were "corpses everywhere, children without heads or with their arms cut off, women with their bellies cut open and the intestines torn out."[3]

By the time the smoke had begun to settle and this seemingly endless day had finally concluded, the Maya village at Finca San Francisco had been all but razed from the face of the earth. Most of the town's 60 or so homes had been destroyed by fire. And more than 300 men, women, and children — perhaps as many as 350, the precise number remains uncertain — had been brutally murdered by the Guatemalan army.[4] At least 91 of them were children under the age of 12; the youngest victim, a baby boy named Andres Paíz García, was less than two months old. No more than two score residents of San Francisco survived the massacre. Among them was not a single woman or child.

In the savage deaths of these innocents, buried beside them like some deeply repressed family secret, lie five centuries of unspeakable cruelty against Maya people — a tragic but true historical tale that still cries out to be told, by Maya voices, to the world.

1989

1. *Voices of the Survivors: The Massacre at Finca San Francisco, Guatemala*, Jason Clay, ed. Occasional Paper No. 10, Cambridge, Mass.: Cultural Survival, Inc. and Anthropology Resource Center, September 1983, 40 pp.
2. Stephen Kinzer and Stephen Schlesinger. *Bitter Fruit: The Untold Story of the American Coup in Guatemala*. Garden City, New York: Doubleday & Co., 1981. (See also Richard H. Immerman. *The CIA in Guatemala: The Foreign Policy of Intervention*. Austin: University of Texas Press, 1982.)
3. *Guatemala: The Human Rights Record*, Amnesty International Publications, 1987, p. 62.
4. Jean-Marie Simon. *Guatemala: Eternal Sping, Eternal Tyranny*, New York: W. W. Norton and Co., 1987.

Epilogue

In memoriam:
The names and ages of the dead that have been documented are: Mateo Paíz García (21), Ana Paíz Domingo, Isabela García Silvestre, Isabela Paíz García (20), María Paíz García (12), Angelina Paíz García (10), Mateo Paíz García (7), Federico Paíz García (6), Francisco Paíz García (4), Andres Paíz García (1 month, 20 days), Isabela Paíz Paíz (1 year, 6 months), Francisco Paíz García, Isabela García Marcos, Mateo Paíz Velasco, Isabela Paíz García, Mateo Paíz Ramos, María Ramos Paíz, Angelina Paíz Ramos, María Paíz García, Francisco Paíz García, Juana Paíz García, Isabela Paíz García, Angelina Paíz García, María Paíz García, Francisco Paíz Perez, María Paíz, Angelina Paíz García, Mateo Paíz, Francisco Paíz, Angelina Perez, María Ramos, Pascual Gomez Ramos (35), María García, Mateo Gomez García (13), María Gomez García (12), Bartolo Gomez García (8), Mateo Gomez García (3), Bartolo

Gomez García (3 months), Mateo Gomez R. (26), Juana Ramos R., María Gomez R. (3), Mateo Gomez R. (2 months), Baltasar Gomez R. (22), Eulalia Paíz, María Gomez R. (19), María Gomez R., Mateo Gomez R., Eulalia Marcos M., Pascual Ramos G. (30), Angelina Ramos L., Mateo Ramos L. (11 months), Juana Ramos M. (20), Andres Lucas B. (6 months), Andres Lucas R., Eulalia Paíz (40), Pascual Paíz (19), Isabela Domingo (20), Eulalia Paíz D. (4 months), María Paíz (13), Angelina Paíz (10), Luxas Paíz (9), Isabela Domingo (30), Mateo Domingo R. (10), Francisco Paíz D. (8), Marcos Domingo P. (7), Gaspar Domingo R. (4), Marcos Domingo R. (1 year, 2 months), Diego Lucas R. (40), María Paíz (30), Marcos Lucas P. (12), Francisco Paíz I. (7), Mateo Paíz Lucas (4), Catarina Perez Lucas (3), Marcos Lucas Paíz (1), Mateo Ramos Diego (72), Ana Ramos, Andres Ramos Ramos (36), Catarina García Paíz, Ana Ramos García (10), Juana Ramos García (16), Angelina Ramos García (9), Diego Ramos García (11), Mateo Ramos García (4), Isabel Pérez Ramos (32), Bartolo Ramos Pérez (9), Diego Lucas Pérez (7), Ana Ramos Pérez (6), Catarina Ramos Pérez (2), Mateo Ramos Ramos (S) (32), Juana Lucas Paíz (22), Eulalia Lucas Paíz (11), María Ramos (40), Eulalia Alonso (80), Juana García Silvestre, María Paíz García (7 months), Ana Paíz Ramos (20), Diego Ramos Andres (42), Miguel Lucas (25), Mateo Lucas Paíz, Juana Lucas Paíz, Bartolo Pérez (65), Catarina Ramos (58), Pascual Pérez Ramos (30), María Paíz Domingo, Bartolo Pérez Domingo (13), Catarina Pérez Domingo (12), Juana Pérez Domingo (10), Ana Pérez Ramos (36), Ana Paíz Ramos (14), Catarina Paíz Ramos (12), Lucas Paíz Ramos (7), Bartolo Paíz Ramos (5), Angelina Paíz Ramos (1), Juana Paíz García (25), Bartolo Pérez Paíz (10), Felipe Pérez Paíz (7), Catarina Pérez Paíz (5), Isabel Pérez Paíz (1), Ana Mendoza (30), Bartolo Mendoza Pérez (14), Petrona Pérez Mendoza (7), Catarina Pérez Mendoza (5), Ana Pérez Mendoza (2), Isabela Pérez, Eulalia Andres (9), Ana Gómez Andres (8), Mateo Ramos Paíz S., Lucas Ramos Mendoza, Pascual Ramos Mendoza, María Mendoza, María García Paíz, Pascual Ramos Mendoza, María Ramos García, Mateo Ramos Mendoza, María Ramos Mendoza, Petrona Ramos Mendoza, Pedro Pérez García (25), María Ramos Juan (20), Pascual Paíz Domingo (28), Isabel Paíz Domingo (12), Francisco Paíz Domingo (7), Isabela Paíz Domingo (3), Andres López Paíz, Isabela Lucas, Miguel Lucas (S), Marcos Lucas Mateo, Isabela Ramos L., Mateo Lucas, Juana Lucas, Isabela García, Catarina Lucas, Ana Silvestre Lucas, Angelina Silvestre Lucas, Isabela Silvestre Lucas, Francisco Silvestre Lucas, Miguel Silvestre Lucas, Angelina Santizo, Felipe Silvestre Santizo, Mateo Silvestre Ramos, Francisco Silvestre Ramos, María Ramos, Ana Silvestre Ramos, Francisco Silvestre R., Angelina Silvestre R., María Silvestre R., Pascual Silvestre, Isabela Domingo Paíz, Mateo Ramos Lucas, Marcos Lucas García, Isabela García, Ana Paíz Ramos, Catarina Ramos Lucas, Andres Lucas García, Isabela Paíz García, Miguel Ramos

Lucas (30), María Paíz García (33), Mateo Lucas Paíz (10), Marcos Paíz Lucas (8), Juana Paíz García (7), Pedro Pérez (90), Eulalia Sebastián (80), Pascual Pérez Gómez (30), Catarina Ramos (35), Pedro Pérez Domingo (18), Martin Pérez Domingo (15), Sebastian Pérez Domingo (10), Mateo Pérez Ramos (7), María Paíz Silvestre (42), Baltasar Paíz García (23), Juana Lucas Lucas (18), Miguel García Paíz (5), María García Velasco (12), Diego García Velasco, Pascual Mendoza, Miguel Mendoza, Lucas Mendoza, Petrona Mendoza, Lucas Mendoza, Isabela Ramos, Andres Ramos (61), Mateo Lucas Lucas (35), Angelina Paíz (25), Juana Lucas (12), Magdalena Lucas (11), María Lucas (8), Pascual Lucas (2 months), Magdalena Lucas Miguel (40), Andres Lucas Carmelo (50), Andres Paíz Lucas (30), Mateo Lucas Paíz (10), Catarina Lucas Velasco (30), Domingo Paíz (10), Bartolo García Pérez (70), Petrona Domingo (50), Andres García D. (40), María Silvestre, Miguel García (10), María García Silvestre (8), Bartolo García Silvestre (5), Miguel García Domingo (40), Ana Santizo, Bartolo Santizo (17), Andres Santizo (3), Petrona García S. (10), Bartolo García (15), María García S. (5), Alberto García S. (3), Andres García D. (40), Petrona Lucas (45), María García L. (15), Bartolo García L. (12), Catarina García L. (8), Petrona García L. (7), Lucas García D. (30), Mariz Paíz R. (25), Antonio García (22), Eulalia Paíz R. (20), Pedro Marcos Martín (40), Isabela García (35), Magdalena Marcos G. (9), Juana Marcos G. (6), Juana Marcos G. (3), Juana Martín, Gaspar Marcos M. (23), Isabela Pérez (23), Juana Marcos R. (2 months), María Gomez Andres (22), Angelina Marcos (2), Eulalia Paíz (50), Angelina Domingo P. (12), Jorge Santizo (75), María García (48), Gaspar Santizo G. (27), Juana Diego D. (23), Juan Santizo D. (5), Jorge Santizo D. (1), Jorge Santizo D. (18), Francisco Paíz, Juana Lucas, Eulalia Ignacio, Pascual Ramos P., María Lucas, Angelina Ramos L., Marcos Ramos L., Mateo Ramos P., Catarina Ramos L., Marcos Lucas (40), Isabela García S., Mateo Gómez S. (14), Marcos Gómez S. (4), Catarina García P. (1 year, 6 months), Pedro Pérez G., Andres Paíz D., María Lucas G. Eulalia Paíz, Miguel Paíz L., Francisco Paíz, Isabela Paíz, Catarina Paíz, Catarina Lucas R., María Pérez, Pedro Ramos Pérez (19), Mateo Ramos Pérez (26), Pedro Ramos Pérez S. (13), Bartolo Ignacio, Isabela Pérez, Francisco Pérez I., María Ignacio P., Ana García P., Miguel Paíz L., María Pérez R., Mateo Paíz R., Juana Paíz R., Catarina Paíz R., Lucas Paíz R., Pedro Gómez P. (35), Angelina Paíz Silvestre, Mateo Gómez P. (12), Gaspar Gómez G. (10), René Gómez G. (5), Mateo Gómez G. (2), Isabela Ramos

Suggested Reading:
For readers who wish to learn more about the Maya's plight, the single most compelling work on the lives of contemporary Maya to date is probably the extraordinary oral history, *I, Rigoberta Menchu — An Indian*

Woman in Guatemala (edited by Elisabeth Burgos-Debray, Verso Editions, 1984) — one Maya woman's vision of the daily terrors being inflicted upon her people inside Guatemala. Eduardo Galeano, the well-known Uraguayan writer, has written an impassioned three-volume popular history of the Americas, including Guatemala, titled *Memories of Fire* (Pantheon Books, 1988). And Jonathan Fried and his co-authors have written *Guatemala in Rebellion: Unfinished History* (Grove Press, 1983) — a scorching exposé of some of the harsh political realities of life in Guatemala. Other important recent publications include Beatriz Manz's *Refugees of a Hidden War* (State University Press of New York, 1988) — a scholarly treatise on the massive displacement of Guatemalan civilians as a result of the military's counter-insurgency campaigns over the past decade. Jean-Marie Simon's *Guatemala: Eternal Spring — Eternal Tyranny* (W. W. Norton and Co, 1987), is a timeless and visually eloquent photo-essay on the paradox of Guatemala's beauty and brutality. And *Harvest of Violence* (University of Oklahoma Press, 1988), edited by Maya scholar Robert Carmack, offers a valuable record of recent political persecution of the Maya in Guatemala in the form of an anthology of reports by respected anthropologists and other academicians. Finally, Canadian Ronald Wright has written a fine travelogue, *Time among the Maya* (Viking, 1989) that provides vivid glimpses into the Maya experience — past and present.

Rushton on Race

Science is an attempt to free investigations and knowledge from human emotional attributes. . . . This so-called "objective world" thus becomes devoid of mind and soul, of joy and sadness, of desire and hope.

François Jacob,
French molecular biologist and Nobel laureate,
The Possible and the Actual

In science the quest for knowledge can often become sidetracked for a time by seductive appeals to simple "absolute truths," which seem to render some of the bewildering complexities of our cosmos instantly comprehensible. Of these, perhaps no siren call has proved more alluring than those scientific theories purporting to explain once and for all the biological basis of vital anatomical and behavioral differences between groups of human beings. This is especially the case when the focus of such theories shifts, sometimes ominously, from suggesting strict genetic limits to the capacities of *individuals* to "proving limits on the capacities of entire ethnic groups or races — prophesying, in essence, their biological destinies.

Canadians were able to witness this process at first hand in early 1989, during the fierce controversy that arose over scientific claims of possible hereditary race-based differences in intelligence, sexual behavior, and other value-laden human traits by a previously obscure young professor of psychology from the University of Western Ontario (UWO) by the name of J. Philippe Rushton.

ON JANUARY 19, 1989, in the Sausalito Room of the San Francisco Hilton Hotel, Dr. Philippe Rushton stood before a lectern, speaking to a symposium made up of approximately 100 scientists and 50 reporters from across North America at the annual convention of scientists belonging to the prestigious American Association for the Advancement of Science (AAAS). The title of the brief paper he proceeded to present to the meeting was "Evolutionary Biology and Heritable Traits (With Reference to Oriental-White-Black Differences)."

It was the subtitle, of course, discreetly set off by parentheses, that would generate much of the subsequent furor. After adjusting his large, dark-rimmed spectacles, the slight, impeccably dressed British-born Canadian psychologist proceeded to offer a carefully worded but astonishing proposition. All humankind, which he lumped into three rather unwieldy racial groups — Mongoloid (i.e., Orientals), Caucasoid (i.e., whites), and Negroid (i.e., blacks) — seemed to fit neatly, and repeatedly, he said, into the same orderly 1-2-3 ranking when he compared them on the basis of a list of more than 50 different human traits.

Rushton's list, as he must have known, constituted an absolute minefield of political controversy. Among the behavioral traits he selected for his racial comparisons were such vaguely defined and elusive human attributes as intelligence, mental stability, law-abidingness, and — even in this era of an explosive global epidemic of AIDS — sexual restraint. The anatomical traits he chose to focus upon were no less disturbing; they ranged from brain size and development of secondary sexual characteristics such as buttocks and breasts to the very dimensions of male and female genitalia.

For each trait, Rushton claimed, the statistical average of Caucasoid measures fell unfailingly between Mongoloids and Negroids. Or, to use Rushton's own clipped and categorical shorthand: *Oriental* > *White* > *Black* (with the mathematical symbol > signifying "greater than"). More important, he contended that this orderly tri-level hierarchy of races in his hypothesis had its roots not simply in a host of contemporary economic, cultural, familial, and other *environmental* forces but, to a far greater extent than mainstream social science would suggest, in ancient, gene-mediated *evolutionary* ones. Heredity, or *nature* — to use the term popularized by Francis Galton, the nineteenth-century British scientist who helped lay the foundation for eugenics, the pseudo-scientific movement dedicated to the improvement of human species through the deliberate control of hereditary factors — was, in Rushton's grand scheme, every bit as important as environment or *nurture*, often more so.

The origin of the stratified racial differences he claimed to have tallied, said Rushton, lay in part in the similarly ordered time at which each of his three groupings branched off from the main trunk of the human evolutionary tree. "One theoretical possibility," ventured Rushton, "is that evolution is progressive and that some populations are more advanced than others." This view is, unfortunately, in open defiance of one of the basic tenets of modern evolutionary theory, which views all living species as well as their constituent subgroupings as, in essence, *equivalent* cases of time-tested evolutionary success.

In Rushton's mind, his hypothesis had global implications. If his evolutionary-model-for-racial-differences hierarchy turned out to be scientifically correct, he said, "then two important predictions can be made about the course of world history." First, so-called evolutionarily favored Oriental populations — the top-ranked race in Rushton's neat hierarchy — could be expected to "outdistance the predominantly Caucasian populations of North America and Western Europe." Here, as elsewhere in his arguments, he made no mention of the fact that one might with equal confidence make the same "prediction" on the basis of a variety of social, economic, or historical factors, ones far removed from natural selection's endless evolutionary siftings of human genes. Second, peoples of African ancestry — relegated to a dismal third place on account of reports he cited of their supposed statistical bent toward sexual promiscuity, social disorder, and criminality, among other things — appeared, he said, to be "especially at risk" for infection by the deadly, blood-borne HIV virus responsible for AIDS. Why? "Most obviously," stated Rushton boldly, "because they will be drawn to have sexual intercourse with more numerous partners, even when they are married."

Not surprisingly, Philippe Rushton and his novel evolutionary hypothesis for racial differences created an instant firestorm of furious criticism. He was promptly challenged from the floor by angry questioners from the audience. And so stunned by his racial bombshells were AAAS organizers that they quickly called a press conference to publicly disassociate the prestigious assembly of scientists from his recent remarks. Rushton himself was utterly unfazed by the response, according to his hometown newspaper, *The London Free Press*, among others. "It's not racist. It's just a fact," stated Rushton coolly, from the eye of the growing media hurricane he had so calmly touched off in San Francisco. "This is my scientific judgment. I have to believe after four and a half years of research that I'm right about racial group differences and everyone else is wrong."

In Canada, media response to the nascent "Rushton controversy" catapulted the unknown professor into the headlines. "Canadian professor provokes uproar with racial theories," trumpeted one in the venerable *Globe and Mail*. "An obstinate professor and his theories of race," thundered another, in *The Toronto Star*. "UWO professor denies race study was racist" read a third from the pages of the local *London Free Press*, which, more persistently than all the others, would continue in the months ahead to nip at Rushton's heels.

AT THE CORE of most criticism of Rushton seemed to be the deeply felt concern that the real message behind his ranking of races was the potentially dangerous notion that race is the *primary determinant of human traits and capacities and that racial differences produce an inherent superiority of a particular race*. Or, worse, that his hypothesis might, intentionally or not, be providing some sort of "scientific" rationale to those in society who actively engaged in *racial prejudice or discrimination*. (Both of these italicized phrases, incidentally, happen to be verbatim definitions of the word *racism* — surely one of the most venomous words in our language — taken from *Webster's Ninth New Collegiate Dictionary*.) Rushton, for his part, has repeatedly denied any allegations that he or his ideas are in any way racist — meaning, by his own definition of the term, which he offered during the course of a series of taped interviews, "treating everybody within a group as if they were all the same." "I do feel that I have been very badly treated." Added Rushton: "I am a serious scholar and I deserve to be treated as a serious scholar."

At the same time that divisions in the public debate over Rushton's views deepened, there were also a few moments of refreshing comic relief. One of Rushton's most vociferous opponents, the London Urban Alliance on Race Relations, decided, literally, to capitalize on the sudden celebrity status of their nemesis. Because they considered his claims concerning the use of anatomical "measures" as reliable indexes to racial differences in intelligence and other traits so patently absurd as to be laughable, they quickly launched a fund-raising campaign based on the sale of custom-designed "Rushton sportswear." There were "Rushton sweatbands" for $6, adorned with conspicuous tape-measure-like markings to parody his astonishing public suggestions that simply by measuring the circumference of people's heads he could reliably predict their relative intelligence. "Rushton T-shirts" for $10, each emblazoned with the tongue-in-cheek motto Rushton U[niversity] — Standard of Excellence. And, for men, "Rushton jogging shorts," complete with an area, calibrated in inches and strategically aligned above the crotch, aimed ostensibly at revealing the true physical dimensions of the wearer's sexual organ — that mystical appendage that seemed strangely to loom so large in Rushton's besieged hypothesis.

Rushton himself seemed, at times, little inclined to try to diminish the furor. No one really expected him to disavow his claims. Nonetheless, there were no heartfelt apologies for the suffering his highly speculative hypothesis might have already inflicted on Canadians belonging especially to the most maligned of his highly judgmental racial groupings. Nor, claimed some scientists, did Rushton seem to display much willingness (fundamental to any genuine dialogue in science) to *modify*, piece by piece, those elements of his new hypothesis that had received the most widespread and withering scientific criticism. This unsettling obstinacy on Rushton's part led one highly regarded academic (who requested anonymity) to describe the controversial psychologist wearily as less a true scientist than a "pamphleteer." And another to refer to the utterly exceptionless 1-2-3 order of Rushton's racial rankings of so many different life-history criteria as totally unreflective of the true complexity of the natural world and, in the end, nothing less than a "parody of the properly quantitative work of science."

Even more disturbing, Rushton sometimes seemed to be actually trying to fan the flames of public controversy by aggressively pursuing further media exposure rather than simply waiting for his peers to respond fully to his politically explosive hypothesis on race. On March 8, 1989, for instance, millions of TV viewers across North America who happened to tune to the popular daytime NBC talk show "Geraldo" were greeted by the image of Rushton pontificating on matters of race. The show's title captured tabloid-TV's vision of Rushton's extremely speculative evolutionary model of racial differences, reducing it to a single, familiar, and gut-wrenching question: "Is there a master race?"

Exactly *why* Rushton selected the rather infamous "Geraldo" show as a scientific forum in which to continue public debate remains something of a mystery. It was hosted by Geraldo Rivera, well known throughout North America for his headline-grabbing TV style, including, for example, having his nose fractured on the show during a "live" brawl that erupted on stage between a black U.S. civil rights leader and a gang of tough, acid-tongued neo-Nazi thugs. In retrospect, says Rushton, his decision to appear on "Geraldo" was "naive." But he consented to appear on the show, he says, because he had become convinced that it was virtually impossible to get balanced coverage of his hypothesis in the Canadian media, and thus he viewed "Geraldo" as perhaps his last chance to publicly air his ideas. Because he personally watched little TV, he had, he insists, simply been unaware of Geraldo Rivera's reputation as a media "sensationalist."

On "Geraldo," Rushton was joined by a panel of experts on race, racism, and the history of science, each one of them adamantly opposed to his racial hypothesis. In his defense, beamed into the New York studio by satellite feed from California, was one of the most vocal advocates of the supposed biological inferiority of black Americans in the land, the aged Nobel laureate physicist (for his co-discovery of the transistor some 40 years before) William Shockley.

An excerpt from one early exchange sets the stage for the explosive debate that would follow:

RIVERA: Dr. Rushton, when measuring intellectual capability among the races, what did you find?

RUSHTON: I found that around the world, Oriental populations scored higher on IQ tests than whites, and the whites scored higher than blacks.

RIVERA: Orientals, whites, blacks.

RUSHTON: Yes.

RIVERA: In your studies, what did you find about criminal behavior?

RUSHTON: With law abidingness, Oriental populations tend to be less criminal than white populations, who are more law abiding than black populations.

RIVERA: Again, Oriental, white, black.

RUSHTON: Yes.

RIVERA: What did you find, sir, about sexual restraint?

RUSHTON: In sexual restraint, Oriental populations are more restrained than white populations, who are more restrained than black populations.

RIVERA: Oriental, white, black.

RUSHTON: Yes.

RIVERA (turning to his studio audience): Is this science or is this racism . . . ? On the basis of sex and brains and brawn, is there a master race?

With characteristic calm, Rushton tried to summarize some of his arguments that hereditary forces were powerfully at play in reported statistical differences between the races for factors as diverse as intelligence and frequency of child abuse to penis size and frequency of orgasm. It was vintage "Geraldo" material. But Rushton's critics also managed to get in their licks. Dr. Charles

King, a black civil rights leader from Atlanta, lashed out at Rushton's views on "my supposed inferiority" and demanded that they be met not with polite scholarly responses but rather "with our indignation, our anger." Barry Mehler, a science historian, offered his own brief, unflinching assessment of Rushton and his scientific theorizing on race. "My Ph.D. is in the field of academic racism. So I'm an expert at recognizing a racist," said Mehler. "Dr. Rushton is a racist." For his part, Shockley (who, as fate would have it, would outlive the airing of this "Geraldo" show by no more than a matter of months) was no less blunt in rebuttal. "My research," he stated grandly, although he was hardly a trained psychologist, "leads me inescapably to the opinion that the major cause of the American Negro's statistical, intellectual, and social deficits are hereditary and racially genetic in origin, and thus not remediable to a major degree by practical improvements in the environment." But reduced by the very electronic technologies his own genuine scientific achievements had helped to create, the crotchety old physicist seemed no more than a blustery "talking head" on the nearby TV monitor. And his outbursts and frantic hand gestures seemed to do little more than inspire ripples of nervous laughter through the studio audience.

The panelist Jerry Hirsch, a respected American behavioral geneticist, characterized Rushton's ideas as nothing more than "bad science." He also offered a telling comment that in the hot, adversarial atmosphere of this "Geraldo" show was all but lost to most viewers. Quoting from a statement issued by the former president of the prestigious U.S. National Academy of Sciences (originally, as a public condemnation of Shockley and Arthur Jensen, a psychologist from the University of California at Berkeley, for their bitter, decades-long advocacy of the widely discredited notion that American blacks were *genetically* inferior to whites in intelligence), Hirsch offered these disquieting words: "If the *scientific community* will not unfrock the charlatans, then science and the nation will suffer."

THE "GERALDO" APPEARANCE was hardly Rushton's final moment in the media spotlight. The event did serve to demonstrate to many that the professor was, like some fellow scientists, an exceptionally driven individual. But he seemed intent on far more than simply arm-wrestling with scientific colleagues over the validity of his academic arguments. Rather he seemed bent on taking on the mantle of a vigorous public *advocate* of his politically provocative ideas long before they had been demonstrated to have any scientific merit within the arena of legitimate science.

This assessment was deeply reinforced in my own mind when I first learned that Rushton had, in the summer of 1987, despite his repeated public claims that he had "*no* policy recommendations to make," knowingly published his evolutionary views (albeit largely on human altruism and other of their more benign aspects) in the rather notorious American "anthropology" publication *Mankind Quarterly*. In his article, Rushton speculates that "ethnic conflict and rivalry" (which he refers to as "one of the great themes of historical and contemporary society") may be rooted in the evolutionary impacts of individuals from groups "giving preferential treatment to genetically similar others." In addition, he proposes that a person's political orientation might also be genetically determined. Writes Rushton, "The makeup of a gene pool [i.e., a human population's total reservoir of alternative genes] causally affects the probability of any particular ideology being adopted." Furthermore, he continues, "religious, political, and other ideological battles may become as heated as they do because they have implications for genetic fitness."

What is important in this case, I think, is not Rushton's views but rather the vehicle he chose to express them. *Mankind Quarterly*, established in 1960, has a reputation for fostering strong ideological themes of its own, including its perennial promotion of eugenics policies and notions of racial "purity." Thus, simply by publishing in *Mankind Quarterly*, Rushton seems, despite his protests, to be at least implicitly promoting a particular policy agenda. During my own brief perusal of back issues of the journal, I encountered, for example, articles on topics ranging from immigration control, Third World sterilization programs, California sperm banks for the intellectual elite (whose donors have included William Shockley) to odd musings on the evolutionary origins of human "prejudice" or the prospects of global clashes between races. My impression is not an isolated one. Robert Proctor of the New School for Social Research in New York, for instance, is quoted in *The Toronto Star* as referring to *Mankind Quarterly* as "one of three of the most far-right and extremist journals now being published in the world."

Over the months, there would also be other news and revelations about the quixotic professor from London, Ontario. There was, for instance, the unsettling news that Rushton and a fellow UWO psychologist, Anthony Vernon, had together received a total of $206,551 (U.S.) between 1984 and 1987 from the Pioneer Fund, a New York–based endowment fund with a slightly dubious scientific reputation. It was founded in 1937 during a pre–Second World War era when the Nazis' genocidal policies had not yet transformed the term *eugenics* from a scientifically fashionable word in many circles to one with profoundly negative connotations. Its original mandate had been to promote "racial betterment" and to educate Americans about the promise of eugenics. In recent years, it has funded, among other projects (several quite respectable), the race/IQ work of Arthur Jensen and William Shockley.

There were also repeated calls by human rights groups as well as several prominent public figures such as David Peterson, then premier of Ontario, and the science broadcaster David Suzuki (the latter during a dramatic televised debate with Rushton at UWO) for the university to fire Rushton on the grounds of academic incompetence and irresponsibility. "The University deplores bigotry, intolerance, and racism in any form," responded UWO President George Pederson, in the campus publication *Western News*. But, he continued, in defense of his beleaguered colleague, "to abrogate academic freedom would be to invoke those very attitudes which the principle of academic freedom itself rejects." Many UWO students disagreed and openly challenged the administration's moral priorities, on a campus not known for its outspoken defense of the rights of racial minorities, by staging angry, anti-Rushton demonstrations on campus. Later Rushton would receive a mild official reprimand, along with certain restrictions on his teaching duties, not for his academic performance but for his failure to comply with strict university ethics guidelines in his own research programs using university undergraduates.

By June, many Canadians were stung again to learn that Rushton had published still another incarnation of his racial hypothesis. This one was basically an elaboration of his earlier speculations, voiced briefly at the AAAS meeting in San Francisco, on race-based "genetic" differences in "susceptibility" to AIDS. Confounding many of his critics, who often took obvious pleasure in dismissing Rushton's racial writings as simply "not science," this hotly controversial item turned up in a respectable, peer-reviewed British journal called *Social Science and Medicine*. (And according to Rushton his AIDS manuscript had passed scientific muster by surviving preliminary review by three experts — not once, but twice.)

In the article Rushton and his co-author, Anthony Bogaert, a former graduate student at UWO, argue that evolutionary history has somehow genetically predisposed "members of some populations" (e.g., blacks) to indulge in activities "that put them at risk for AIDS." Any implication that biological differences in race might have played a fundamental role in determining the global epidemiological pattern of the AIDS pandemic was of course simply absurd. The pattern of distribution of AIDS among the world's peoples has its roots in the geographical origins, or epicenters, of infection, as well as in such factors as vast inequalities in living standards, immune status, and health care. Robert Gallo, a world-renowned American AIDS expert, promptly dismissed Rushton's claims as "absolutely silly." And Alastair Clayton, director of Canada's Federal Centre for AIDS in Ottawa, concurred. "There is no evidence," said Clayton in an interview quoted in *The London Free Press*, "that AIDS is any way influenced by racial or genetic susceptibility. What we do know — and I think that Professor Rushton is missing the point — is that AIDS is caused by a virus."

Finally, in another important development, the pornography and hate literature squad of the Ontario Provincial Police (OPP) launched a thorough criminal investigation of Rushton, following complaints that his views constituted hate literature. This had the effect of at least temporarily silencing much of the public debate about Rushton, adding to the widespread illusion that the "disappearance" of the Rushton story from the Canadian media had somehow signaled that the issues it has raised had disappeared too.

By the time I finally arranged to interview Rushton over a period of two days in the late summer of 1989 in London, Ontario, he had, he said, declined virtually all interview requests for approximately the preceding four months. In November 1989, the OPP announced the decision that they would not lay criminal charges against him. If they had, Rushton told me defiantly months earlier (in reference to the famous 1925 U.S. court case over the right to teach evolution in the classroom), "it would be the Scopes monkey trial *all over again*."

After this deluge of information on Rushton and his ideas, many Canadians dismissed his statements on race as pseudoscience. Others, I have found, are genuinely confused by some of the technical debate and scientific half-truths that seem to still circulate in the wake of public debate. And some, not surprisingly, appear to have found in Rushton's race-based evolutionary "ladder" convenient "scientific" confirmation of their own intuitive hatreds and fears, some of them blatantly racist, concerning many visible and invisible human differences. According to the results of the Gallup poll, announced in *The Toronto Star* on March 13, 1989, fully 13 percent of all Canadians were willing to admit they were in agreement with Rushton's unproved proposition that "Orientals are genetically superior to whites." (The percentage was highest among respondents with the least number of years of formal education.) Seventy-three percent disagreed with him; 14 percent simply did not know.

Who, after all, many seemed to be asking, was this Philippe Rushton and what exactly *was* he saying? Why *had* David Suzuki, arguably Canada's most trusted demystifyer of science, seemed so visibly *outraged* by the mild-mannered psychologist's "monstrous claims" during their packed debate on the UWO campus? ("There will always be Rushtons in science," Suzuki had thundered, "and we must always be prepared to root them out!") And, finally, why weren't more scientists like Suzuki stepping into the fray? A brief article in the June 1989 *Saturday Night* by the editor, John Fraser, reflects, I think, the legitimate sentiment among many Canadians that the controversies swirling around Rushton are still far from resolved. "The truth is," wrote Fraser "that we have been left to our own resources in sorting out the possibly conflicting requirements of truth and tolerance, of freedom of expression and incitement to prejudice, of harmonious relations and cleaving to principle."

JOHN PHILIPPE RUSHTON was born on December 3, 1943, in Hampshire, England. His parents (his father owned a construction business), he says, were quite "ordinary," and "solidly middle-class" people. In 1948, when he was only four, he immigrated with his family to South Africa, where he lived and studied until the age of eight in an "all white, ethnically mixed" school (i.e., including Afrikaans-speaking Dutch descendants and English-speakers but not South African blacks). Following this brief, post–Second World War sojourn outside the British Isles in search of greener economic pastures, the Rushtons returned to England, where Philippe apparently lived and continued his studies well into early adulthood.

Beyond this, the details of Rushton's family life and youthful experiences in Europe and South Africa are scant, in large measure because of his extremely guarded response to questions concerning his personal history. When in a brief phone conversation I first broached the subject with him — in particular, the obvious question regarding any possible influence his early years in racist South Africa might have had on his choice of research topics — I was momentarily startled by his irate response. Pursuing such "Freudian sorts of notions," he admonished me, could only prove "sterile." Such questions, he huffed, tended to be "either misguided or prurient."

In 1970, at the age of 27, Rushton earned a baccalaureate degree in psychology with first-class honours from the University of London. During his undergraduate studies, Rushton says, he was not dissatisfied with the prevailing scientific faith in the environment as the major determinant of human behavioral differences. But today, looking back on the lack of emphasis on genetics and evolutionary theory in his undergraduate courses in psychology, he says with a trace of bitterness, "I was misled by those professors. They lied to me; they turned me away from the true path."

In 1973 Rushton was granted his doctorate in social psychology from the London School of Economics and Political Science. Ironically his Ph.D. thesis was a study of factors shaping the development of "generosity" in seven- to 11-year-old children. It reflected the strong emphasis he would place on *environmental* (as opposed to genetic) factors underlying various forms of human selflessness, or altruism — his primary research interest during the early years of his career.

Following his doctoral studies at the University of London,

Rushton carried out post-doctoral research in psychology at Oxford. By 1974 he had assumed the post of a visiting assistant professor at Toronto's York University. He would hold similar positions at both the University of Toronto and University of Western Ontario in subsequent years, before beginning, in 1977, the faculty appointment at UWO that he holds today.

By Rushton's own account it was between 1980 and 1983, after he had written a book on social learning and altruism, that his reading of data from studies of the heritability of behavior in human twins compelled him to undergo what he refers to as a major "paradigm shift" as a psychologist. This transformation from "the prevailing orthodoxy of social science," as he describes it, represented a pivotal moment in his career as a social scientist and perhaps the first stirrings of his emerging racial hypothesis. It also resulted in his abandonment of many of his long-held views on the environmental (i.e., *nurture*-oriented) origins of human behavior in favor of a decidedly more hereditarian (i.e., *nature*-oriented) one, in which, he has written, "traits of character" such as altruism were seen as "rooted in the genetic makeup of the individual."

It is worth noting that Rushton took leave from UWO in 1981 — in the midst of his "paradigm shift" — to serve as a visiting scholar at the University of California at Berkeley. There he developed a lasting intellectual bond with the educational psychologist Arthur Jensen who would exert so important an influence on Rushton's thinking that Rushton today points to him as his "mentor." "I am very proud," says Rushton, to be considered "a follower of Arthur Jensen."

It was Arthur Jensen, in 1969 (as I well recall from my own undergraduate days on another, more southerly, campus of the University of California), who had set the scientific community momentarily abuzz with his controversial article titled "How Much Can We Boost IQ and Scholarly Achievement?" published in the *Harvard Educational Review*. In it, he argued *against* the prevailing opinion that environmental disadvantages were responsible for the reported statistical disparity in average IQ scores — about 15 points — between black and white Americans. "The possible importance of genetic factors in racial and behavioral differences has been greatly ignored," wrote Jensen, "almost to the point of being a tabooed subject." That bias should not be allowed to continue, he contended, for it was now reasonable "to hypothesize that genetic factors may play a part" in racial differences in IQ scores.

It was an idea Philippe Rushton would echo and greatly expand upon precisely two decades later from the opposite shore of San Francisco Bay. The historical parallels between the meteoric rise, exactly 20 years apart, of these two bright, Caucasian psychologists with so much to say about the possible relationship of genes, IQ, and race stirred up in many Rushton watchers a distinct sense of *déjà vu*. And this feeling was also shared, not surprisingly, by at least one of Rushton's staunchest defenders. "The backlash that Phil got," said his UWO colleague Anthony Vernon, "was akin to what Jensen got in 1969 in his *Harvard Educational Review* article, when he first proposed that there might be a genetic basis to intelligence differences between races." The noted science historian Daniel Kevles recently offered his expert assessment of that same paper, seen through the rearview mirror of the history of eugenics, in his landmark book, published in 1985, *In the Name of Eugenics*. "No single publication did more to precipitate the revival," wrote Kevles, "of attention given to the issue of race and intelligence" than did Jensen's 1969 article in *Harvard Educational Review*.

In 1983, Rushton was also on a sabbatical with the well-known British psychologist Hans Eysenck, author of such books as *The IQ Argument — Race, Intelligence and Education* (1971) and *The Inequality of Man* (1973). Kevles describes Hans Eysenck as "the most prominent champion of hereditarianism in England." In recent years, Eysenck has edited an international journal called *Personality and Individual Differences*. It appears to have published more of Rushton's controversial articles on possible evolutionary differences between the races (sometimes redundantly) than any other single journal in which Rushton has published in the past several years. It is worth mentioning that unlike the majority of science's most highly regarded scientific journals, it often publishes such papers by Rushton and others without the laborious process of preliminary review by teams of disinterested scientific referees.

Despite the occasional media stereotype of Rushton as some sort of incompetent scientific adventurist, he has throughout most of his career as a psychologist been seen as a highly competent researcher. He has published more than 100 papers, most of them, particularly those dealing with altruism, in highly respectable journals.

Rushton has also written a number of scholarly books, including a pair on altruism, one on "scientific excellence," and a psychology text (co-authored). In 1988 (*after* his first several publications on race), he was awarded a John Simon Guggenheim Fellowship, the prestigious American award given to promising scholars "who have demonstrated exceptional capacity for productive scholarship." Not surprisingly, grateful acknowledgment for research support from the Guggenheim award appears in a number of Rushton's most controversial papers on race. At one point Rushton sent me a narrative account of his research career that he had written in support of his application for the 1988 fellowship. I was surprised to discover, after scanning his succinct three-page account of his research interests (including a discussion of his evolutionary hypothesis on human differences) that *nowhere* could I find a single mention of that sometimes contentious word "race."

At this writing, Rushton is on sabbatical leave, in London, from his teaching duties at UWO. He is busy, he says, responding to fresh volleys of scientific criticism of his racial hypothesis, mostly with brief letters and commentaries. For those among his critics who are convinced that Rushton's recent absence from the media signals the end of his meteoric rise on the Canadian science scene in 1989, it is probably worth mentioning that he is now in the process of writing another book, this one on contract for Cambridge University Press. It will, he told me, represent the culmination of *all* that he has learned to date about the roots of human differences, including, of course, his views on the importance of race.

As WE SIPPED coffee in the stately surroundings of the University of Western Ontario's faculty lounge, where we met for one of a series of interviews in summer 1989, amid elegant maroon armchairs, tan carpets, and heavy polished wood coffee tables, Philippe Rushton told me he was quite pleased to be "following in the footsteps" of a line of scholars that began about a century ago with the British eugenicist Francis Galton. Rushton also insisted on including Galton's more illustrious cousin, Charles Darwin, in this family tree. Galton was best known for his sophisticated statistical research and rankings of human performance and his belief that many of the most valued human qualities are inherited, as suggested by the title of one of his better-known books, *Hereditary Genius*.

This scholarly pedigree, Rushton says, surfaces in the early twentieth century in the names of some noteworthy psychologists, a conspicuous number of them, like him, born and educated in Great Britain. Among those whom Rushton mentions are Charles Spearman, who came up with the idea, still hotly disputed, that human intelligence can be reliably measured by IQ tests because they possess a common, if elusive, component of intelligence he dubbed "g." And there is Sir Cyril Burt, famous for championing widespread use of IQ testing in England so young students could be channeled into educational programs that matched the "innate" differences it monitored. Burt was later infamous for, in the blunt assessment of the noted Harvard science historian Stephen Jay Gould in his award-winning book, published in 1981, on scientific racism called *The Mismeasure of Man*, his "documented fabrication of data on IQ of identical twins." Today, Rushton says, this tradition has found expression in the theories of Hans Eysenck in Britain, Arthur Jensen (who once studied under Eysenck and is "probably a genius," says Rushton glowingly) in the U.S., and Canadians like UWO's Anthony Vernon (who carried out doctoral studies under Jensen) and Rushton, among others.

In conversation Rushton has an animated manner. His dark hair is parted neatly, and his attire is meticulous: a casual summer sports jacket, neatly pressed shirt and slacks, and canvas loafers, all niftily color-coordinated in pale gray. His large, plastic-rimmed glasses dominate his somber face, giving him a certain exaggerated scholarly appearance. With his face and conservatively clean-cut look, I can't help noticing, with an inward smile, that he *does* bear some likeness to the bespectacled Clark Kent caricature that the media have chosen for him. But there is also something missing. Rather than the square jaw and steely gaze of some Superman figure, what is most striking about Rushton is his apparent fragility. He is slightly built, with pale, almost translucent skin, a prominent nose, and dark, slightly weary eyes narrowly set behind the large spectacles. And his thick, expressive eyebrows seem sometimes to arch quizzically after a question, lending him an air, paradoxically, of almost childlike bewilderment.

But Rushton's mild-mannered demeanor reflects no lack of confidence. He tends to speak about his racial hypothesis with a sense of absolute conviction that sometimes borders on romanticism. He is fond of comparing his own lonely public stand on the evolution of race differences to the great Italian astronomer Galileo's courageous defense, to the Inquisition in the seventeenth century, of the Copernican notion of a sun-centered system of planetary motion — eventually, of course, vindicated by science. "It has to do with science; it has to do with absolute knowledge," Rushton said, leaning forward excitedly to emphasize his words. "I mean, you can tie Copernicus to the stake," he said, momentarily confusing his heroes, "but that doesn't change the *facts.* Even if I were to stop my work and do something else, that doesn't change the *facts.*"

That confidence also extends to Rushton's assessment of what he considers to be his own singular contribution to the science of human differences. When I asked him to summarize what he views as the true significance of his racial hypothesis, he replied proudly, "Mine is the first scientific theory of the evolution of racial differences in behavior." Questioned further, he added, "*Nobody*, as far as I know, applies population biology to human differences. *Period.* I don't know of anyone else who is doing what I'm doing."

In this last assertion, Rushton is not of course literally correct. The entire discipline known as sociobiology, for example, is based on the systematic exploration of evolutionary differences in populations — human and nonhuman. In his Pulitzer Prize–winning book, *On Human Nature*, published in 1978, Edward O. Wilson, the distinguished Harvard entomologist and one of the field's most illustrious pioneers, describes sociobiology as "the extension of population biology and evolutionary theory to social organization," and its ultimate aim as "the uncompromising application of evolutionary theory to all aspects of human existence" (including, of course, many of the same sorts of traits Rushton is studying).

But Rushton *is* justified in drawing attention to his rather exceptional willingness to focus so exclusively on matters of race. For the most part mainstream sociobiology, during more than a decade of bitter battles over all sorts of mainstream issues, has largely refrained from applying its evolutionary analyses to possible racial differences in intelligence and other value-laden human qualities. In this, E. O. Wilson himself has been a model of scientific decorum. He has, to be sure, stirred up his own share of controversy for openly speculating, for example, on the possible evolutionary roots of everything in human society from gender roles to homosexuality. (He has even, I was surprised to discover, publicly pondered the possible "evolutionary value" of genocide, which, as he wrote with absolutely stunning scientific dispassion in his book *Sociobiology: The New Synthesis*, possesses "the interesting property" that it "need take place only once every few generations to direct evolution.") But Wilson has also, it must be emphasized, won international respect for the admirable scientific rigor of much of his evolutionary analysis, and — unlike Rushton, whom he knows personally — has publicly managed to carefully sidestep, for the most part, sociobiological topics concerning race.

In his book *The Biology of Race*, the biologist James King offered this insightful observation: in the entire index to Wilson's classic *Sociobiology* text, the term "race" does not appear even once. When I asked Wilson why, his reply was unequivocal: "I, like most people, wish to be *extremely* cautious about the public airing of scientific opinions that impinge in any way on minority status or racial variations. I think we have to be *super* careful in that regard. You can quote me on that!" And, after reviewing references to his ideas in a draft of this article, he eloquently underscored his misgivings about such public debate by comparing the challenge of my journalistic endeavor to "trying to examine the inside of a powder keg with lighted matches."

Obviously, I don't entirely agree with Wilson on this matter. Once a mischievous argument like Rushton's has been publicly aired, it must, I believe, be promptly and thoroughly countered in public — especially by leading members of the scientific community. Otherwise a trail of potentially dangerous half-truths and oversimplifications may linger.

UNFORTUNATELY, IT IS impossible, in this fairly limited space, to even attempt to do justice to the full range of scientific criticism, most of it fairly inaccessible to the general public, that has so far been leveled at Rushton's arguments for an evolutionary basis to racial differences in behavior. Nor, as I found repeatedly, are all those scientists who are both expert in fields crucial to Rushton's racial hypothesis and familiar with the intimate details of his arguments always willing to submit their counterarguments for publication or the public record. Still, it is important for ordinary Canadians to have a sense of just how *fundamental* some of the scientific challenges to Rushton's hypothesis have been.

Interestingly the first systematic attack on Rushton's views on race appeared in 1988 — well before the January 1989 AAAS meeting in San Francisco that turned Rushton's name into a virtual household word in Canada. Rushton, by his own account, had published aspects of his racial model as early as 1985. In other words, the public outcry following the AAAS presentation simply represented the first widespread recognition by the media of Rushton's racial message. "In some ways," Rushton told me, "I was surprised it hadn't happened earlier. I wasn't prepared for it happening at AAAS especially. And I certainly would never have predicted the depth and the thoroughness of the opposition. That has certainly taken me aback. It's limiting. It's censorship."

This brief rebuttal to Rushton, eyecatchingly titled "Oysters, Rabbits and People: A Critique of Race Differences in Behavior by J. P. Rushton" was written by the American psychologists Marvin Zuckerman and Nathan Brody and published, commendably, in the same issue of Eysenck's journal, *Personality and Individual Differences*, in which Rushton's evolutionary hypothesis of race differences in behavior appeared. In it, the two authors criticize Rushton's work harshly for a multitude of scientific sins. Among them:

- "obscure logic," including — as Suzuki had emphasized in his debate with Rushton — his claims of a *causal* connection among a host of traits (i.e., IQ scores, penis size, cranial size, sexual practices, and all the rest) and fertility — that great engine of human evolutionary change — when, at best, only a *correlative* one had been shown to exist
- "biased analysis of data"
- excessive selectivity in its review of scientific literature

- questionable "credibility of some of its sources"
- inadequate consideration of the "diversity in subgroups within races that may exceed differences between the races"
- lack of measures of the "statistical and scientific significance" of those behavioral differences between the races he had reported
- failure to adequately acknowledge such potent forces of *nurture* as "racial differences in socioeconomic status," which, few would deny, arise more from environmental influences than from the dictates of DNA

At times, the two psychologists openly mock Rushton for what they view as his transparent methodological sleight of hand. The writings of a "French colonial army surgeon from 1898," they write scornfully, "is hardly a credible and unbiased source of anthropology." (Another Rushton critic, Neil Weiner, a psychologist from York University, went further, according to an article in *The London Free Press*, referring to Rushton's work as "worthless" and "anthropology of the semi-pornographic sort.") Furthermore, they chide Rushton, "It is strange to see the revival of the curious science of craniometry [the comparison of human skull dimensions], a pseudoscience used to prove 19th century theories of racial superiority, despite the glaring contradictions in the data."

One of the major vulnerabilities of Rushton's hypothesis turns out to be the very biological nature of "race." E. O. Wilson, in *Sociobiology*, offers this definition of a race, or subspecies: a "population that differs significantly from other populations belonging to the same species" and that is "separated from other subspecies by distance and geographic barriers that prevent the exchange of individuals." It is the last part of this definition, in particular, Rushton's critics suggest, that causes him serious problems. For in his attempt to pigeonhole all *Homo sapiens'* rich genetic diversity into just three "races" — Mongoloid, Caucasoid, and Negroid — he has vastly oversimplified the complexities of biological reality. Furthermore, human history, they argue, is replete with all sorts of geography-defying minglings between populations, ranging from long, overland migrations and oceanic voyages to generations of between-group matings.

No one is suggesting, of course, that there exists no genuine geographic variation between some human groups, or that there are not groups that display, on occasion, measurable statistical differences in one biological trait or another (in skin pigmentation, for example, or in height). But what does remain unresolved is any final definition of the term "race" for *Homo sapiens*, as well as the precise number of races in existence (which varies from three to more than 100) and their boundaries (because such groups differ principally not so much in the kind of genes they possess as in their relative *frequency*). "Human population biologists know," says Emoke J. E. Szathmary, a physical anthropologist and dean of the UWO School of Social Sciences, "that the number of human races is in itself open to dispute. It all depends on what traits one chooses to define a race, and the choice is dependent on the type of problem one addresses."

Another technical Achilles' heel of Rushton's evolutionary model appears to be his data on the relative antiquity of his three "races." Crucial to Rushton's hypothesis is his contention that human races — *however* one chooses to define them — split off from the ancient ancestral hominid line in neat, sequential order to evolve independently — each, in essence, in its own ancient and distinctive environmental arena. The chronological order in which the three groups diverged from the mainstream of ancestral human beings, says Rushton, coincides exactly with his own hierarchical scorecard for the races. In other words, the neat 1-2-3 pattern in the timing of racial splitting-off he reports (i.e., beginning with the most recently emerged: *Orientals* > *whites* > *blacks*) perfectly matches, he suggests, and is responsible for the "multifarious other differences including brain size and intelligence, rate of maturation, sexuality, personality, and social organization" he claims to have seen.

However, the data Rushton cites in support of his three-step emergence of the races — Negroids first, about 200,000 years ago; then Caucasoids, about 110,000 years ago; finally Mongoloids third, about 41,000 years ago — are simply not as authoritative as he seems to suggest. There are, in fact, perfectly plausible alternative splitting dates — even allowing, strictly for the sake of argument, the validity of his chosen "racial" divisions — that do *not* reinforce his rankings.

In any case, evolutionary theory simply offers no meaningful assessments of the relative "superiority" or "inferiority" of one or another living biological race or subspecies based on the dates when they split off from an ancestral population. Rather it sees every living population of organisms on the planet as a proved evolutionary success that has by its very survival shown itself to be exquisitely adapted to its environment. In a published letter criticizing Rushton's views, Szathmary also says, "As for superiority and inferiority — any geneticist finds this notion amusing. Each is dependent on the environmental context within which the 'superior' or 'inferior' trait (not population) evolved. Change the environment, and one may change the labelling attributed to the trait."

There are also potentially devastating challenges to Rushton's application of certain sophisticated ecological models to buttress his racial hypothesis. In his scientific publications, Rushton often refers to his hierarchical racial hypothesis as his "r/K evolutionary theory of racial differences in behavior." This title may be perfectly acceptable in the jargon-laden community of scientists and social scientists, but it has sometimes proved misleading in public forums. By his rather casual use of the word *theory* (which is more commonly used to refer to widely accepted notions in science), as well as the esoteric ecological symbols "r" and "K," he seems to grant his politically explosive views a scientific legitimacy that they have not earned.

The basic principles surrounding the so-called r/K model of population change that Rushton relies on so heavily appeared first in the literature of ecology more than a quarter of a century ago. They are credited in large measure to the distinguished American population ecologist Robert H. MacArthur, along with his subsequent collaborator, E. O. Wilson. Briefly the "r" and "K" refer to parameters in mathematical equations designed to show changes in population size resulting from evolutionary pressures exerted by a range of possible environmental settings. In a strictly practical sense one might think of r and K as labels for two (though hardly the *only* two) alternative breeding strategies exhibited by some species as they undergo natural selection in their own particular niche, or environmental setting, in the natural world. Simply stated, so-called r-strategy species (such as the dandelion or salmon) could be said to enhance their survival by placing their evolutionary bets on producing vast numbers of offspring, and their population growth tends to be explosive since it is not held in check by environmental limitations. K-strategy species, in contrast (the less prolific elephant or human being, for instance) could be said to invest their limited energy budgets in fewer progeny but more parental care. And their population growth is usually characterized by a state of equilibrium or dynamic balance (known as carrying capacity, or K) that is imposed in much the same way a thermostat regulates temperature, by limitation in the environment.

Because all of us share membership in the species *Homo sapiens* — a classic K-strategist, according to ecologists — every human race, however variable the definition, can legitimately lay claim to a legacy of K-type natural selection. However, Rushton is bent on determining which human groups might be *more* K in terms of their breeding strategies. (He refers to his attempt to monitor a group's "K-ness" as "differential K theory.") But in the process, Rushton's version of the r/K model is, according to some population ecologists, fundamentally flawed and his conclusions simultaneously theoretically invalidated by his use of "inappropriate units of analysis." The argument goes something like this. Strictly speaking, the concept of r vs. K selection in population ecology applies *only* to the evolutionary processes at work within genuine *populations* — that is, groups made up of members of the same species that are also sharing a particular geographical area. Rushton applies the concept to entire human *races*, which are not, strictly speaking, true biological populations at all. Human "races," as discussed earlier, are in fact far more complex and arbitrarily defined groupings because of their high mobility and often elusive geographical boundaries.

His analysis, says Rushton, suggests that Mongoloids are actually "more K-selected than Caucasoids and especially more than Negroids." Moreover, he maintains that differences in race-based reproductive strategies, acted out in a diversity of ancient, evolutionary arenas around the globe, played a central role in creating the unfailingly predictable pattern of racial differences (*Orientals > whites > blacks*) in intelligence, sexual restraint, and the suite of other behaviors that he assumes scientists have successfully managed to statistically measure.

Critics are suggesting that Rushton's evolutionary analysis is also seriously flawed in other ways. They point, for example, to his tendency to simply ignore other non-r/K evolutionary processes that might also be at play. They are also concerned about his inclination to use r/K terminology in a rhetorical rather than strictly scientific sense, without taking time to propose the sorts of research studies that would be necessary to begin to demonstrate — to the extent it can actually ever ethically be done in *Homo sapiens* — whether such evolutionary processes are *really* at work in representative human populations. And they are suspicious of his highly selective use of only those speculative, evolutionary "just-so stories" — the sociobiologist's term for fantastic Kipling-esque tales that can be conjured up to account for virtually any human or animal trait from a leopard's spots to a tortoise's shell — that happen to neatly conform with what seems to be his hierarchical model of racial differences.

Some of the most telling challenges to Rushton's racial hypothesis have focused specifically on the readiness with which he, like many others, so casually makes use of IQ scores as if they were an unquestioned "measure" of human intelligence. Psychology's long twentieth-century struggle to come to terms with the elusiveness of human intelligence and the significance of IQ performance — not to mention the "heritability" of intelligence — are far too complex to try to untangle here. Suffice it to say that although there is a consensus that IQ scores seem to have predictive value in some limited ways (academic performance, for instance), IQ tests by their very nature can be fraught with biases, uncertainties, and limitations. And no one, to my knowledge, has yet come up with a simple, reliable test to "measure" other deeply cherished human attributes such as imagination (an "IQ" test of a different sort!), creativity, compassion, or countless other dimensions of human mental activity that may well have played an evolutionary role in the past, and may do so again in the future.

As to perennial questions regarding the possibility of a genetic basis to racial differences in IQ scores, perhaps the most powerful argument was made almost two decades ago. As David Suzuki emphasized in the Rushton debate, most scientists were persuaded by the well-known conclusions reached by the geneticists Walter Bodmer and Luigi Luca Cavalli-Sforza in a classic 1970 report in *Scientific American*. In it they asserted that it was "practically impossible" to even carry out the sorts of legitimate scientific studies that could conceivably reveal any underlying racial difference in IQ performance. The reason, quite simply: in societies such as the U.S. (or Canada, for that matter), the very *existence* of racism precluded the possibility that a black child could ever be "brought up in a strictly comparable environment" to that of a white child. For now, writes Stephen Jay Gould in *The Mismeasure of Man*, "What craniometry was for the nineteenth century, intelligence testing has become for the twentieth, when it assumes that intelligence (or at least a dominant part of it) is a single, innate, heritable, and measurable thing."

Beyond the issue of the scientific validity of racial IQ claims, of course, there is also the issue of the human value of such research. Even if such research were one day possible and even if unequivocal hereditary differences between races in IQ performance were someday detected, what conceivable good could such information do, compared to its potential for harm, in a society that claims to grant all its citizens equal opportunity for education, employment, and other valued life experiences?

IN A NOISY LITTLE CAFÉ in London, toward the end of several hours of questioning Rushton, while we were waiting for a flat tire on his Volvo to be fixed, I finally simply could not refrain from venting a bit of my own simmering anger over a particularly obnoxious feature of his so-called "scientific" hypothesis about evolutionary differences between races. What had especially incensed me was his casual use, with all that it seems to imply, of his mathematical *Oriental* > *whites* > *blacks*, the recurring and so easily remembered "formula" that was too often all that remained in the public's perception of his intricate web of intellectual racial views. In the face of the staggering level of biological variation known to exist both *within* (where the vast majority of human variation resides) and *between* human groups, what *possible* scientific meaning could it have? More than this, why did he still stubbornly *persist* in publicly flaunting it — long after he had become aware of serious reservations many scientists have about the validity of the highly speculative ranking of races it is meant to reflect?

"It's just a ranking," Rushton replied with a flicker of a smile.

"Then *what exactly* did you intend the formula to mean?" I persisted.

"It refers to the average difference — the mean. The Oriental *mean* for IQ, for example, would be 106, with a standard deviation [a statistical measure of score range] of 10 or 15. The white average would be 100, with a standard deviation of 15. And the black average would be about 85, with a standard deviation of about 10. So it's a *ranking* of mean differences, average differences."

"And there is no implication in that formula of some sort of evolutionary *superiority*?"

"Oh, no!" exclaimed Rushton. "Obviously," he continued, "you might rank a *particular* trait — like intelligence — as a high trait. For sexuality — well, different people have different values."

"But wouldn't you agree that high intelligence tends to be valued fairly universally?"

"Yes, it is," he answered. "But again, from an evolutionary point of view, superiority can only mean adaptive value — if it even means this. And we've got to realize that each of these populations is perfectly, beautifully adapted to their ancestral environments."

Beautifully adapted to "ancestral" environments, but not, Rushton's eloquent yet evasive caveat seemed to imply, to modern environmental conditions — that modern stage upon which living human populations, with all their evolutionary baggage of biological differences, are seen by Rushton and like-minded others to be so unequally competing.

I had occasion to discover that I was not the only one to find that Rushton's simple "rule" on racial differences had, like some banal advertising jingle, been burned into my brain. While they are mere anecdotes, I offer these experiences as illustrations of what I think is one deeply troubling aspect of Rushton's public advocacy of his politically potent but scientifically unfinished ideas. His formula, I found, was the mnemonic device by which a number of Toronto cab drivers with whom I chatted following my interviews with Rushton in London were able to recall the major themes of Rushton's ideas on race. And, one day, to my surprise, during a break from library research for this article, it showed up in the midst of freshly scribbled, blatantly racist graffiti on the bathroom wall of a major Canadian university thousands of kilometers from Rushton's hometown of London, Ontario.

Charles Lumsden, a theoretical sociobiologist at the University of Toronto who knows Rushton and his writing well (and has co-authored an article with him on more benign aspects of the evolution of human social behavior), concurs with my reaction. Rushton's persistent use of this mathematical shorthand, he told me, was misguided. Lumsden is the co-author with E. O. Wilson of a book titled *Genes, Mind, and Culture*. Its sociobiological themes on the possible genetic foundations of many features of modern human societies happen to be pivotal to some of Rushton's arguments about race, but Lumsden, like a number of respected scientists, sees Rushton in a somewhat paradoxical light. He is quick to dismiss what he considers to be Rushton's simplistic three-layered stratification of racial differences as "absolutely preposterous." At the same time he lauds Rushton's efforts as legitimate science — "doing what science can do" — and unabashedly says that "Phil is a fine scientist" and "a scholar." Nonetheless, Lumsden told me, Rushton's seemingly casual use of value-laden mathematical symbols and language in discussing a topic as sensitive as race was a mistake. For the symbol in the context of race could only be described as "emotionally supercharged."

ARE THERE LASTING LESSONS to be learned from the deeply polarized and unfinished debate over Philippe Rushton and his divisive ideas about evolution, behavior, and race? There are, of course, many — too many to try to raise, let alone try to resolve, here. And most of them, I suggest, strike at the very heart of the relationship between science and the society that science serves.

Certainly we need as a society to become much better versed in the lessons that the history of science might offer us (even if not all of us agree on their precise interpretation). Consider, for instance, Rushton's sudden appearance on the Canadian scientific scene in the light of a roughly simultaneous resurgence in public interest throughout North America in immigration policies, multicultural education, and the problem of urban racial violence. Is there a relationship? "The history of scientific views on race," writes Stephen Jay Gould, "serves as mirror of social movements.

This mirror reflects in good times and bad, in periods of belief in equality and in eras of rampant racism. The death knell of the old eugenics in America was sounded more by Hitler's particular use of once-favored arguments for sterilization and racial purification than by advances in genetic knowledge."

Canadians might consider, too, the thumbnail assessment of Rushton's possible role in the pantheon of twentieth-century eugenicists given to me in a phone interview by Barry Mehler, an American science historian and an authority on institutional and academic racism. "I think Rushton is a young man on the way up," Mehler told me. "And I think that what we are seeing now is a new generation of eugenicists. There are a lot of old-timers doing this sort of thing; now you have younger people who are taking the torch to the new generation. Philippe Rushton is one of them. One might say he is riding shotgun for the new eugenics. He is, in a sense, a new lightning rod."

This may help to illuminate the repeated comments I heard from a diversity of concerned Canadian Rushton-watchers, ranging from journalists and human-rights activists to a number of respected academics who preferred that their opinions not be directly quoted. They generally conclude that many of the main themes of the Rushton controversy seem to be a historical rehash of earlier scientific debates over postulated differences between races. "In a sense," said Fredric Weizmann, a psychologist at York University and a publicly outspoken critic of Rushton, "every generation seems to have to relive the debate over these sorts of issues. Don't forget, too, that we are also a long time from the Holocaust and a long way from Hitler. So we can't always draw upon people who possess a direct memory of where some of these kinds of ideas can lead to."

It is also crucial in this matter to maintain proper perspective — to try to distinguish clearly between Rushton the individual human being with his outspoken views and unfathomable motives and the larger, longer-lived, and equally accountable community of scientists and other academics. It is the scientific community, it must be remembered, that has tended, sometimes reluctantly, to embrace Rushton as one of their own — by occasionally single-mindedly shielding him with the cloak of academic freedom and by lending scientific legitimacy to his often wildly speculative evolutionary views on race, IQ, and HIV transmission in reputable academic journals.

It would also be naive to ignore the fact that there exists a measure of scientific sympathy — for the most part publicly silent — for Rushton's evolutionary thinking on race. Charles Lumsden sweepingly referred to this group as an "invisible collection of Rushton's reluctant supporters." Although it may come as a painful surprise to some in the scientific community, one of them happens to be none other than E. O. Wilson (although, to my knowledge, this significant fact has, at this writing, yet to be reported in the Canadian media). "I think Phil is an honest and capable researcher," Wilson told me in a cautiously worded statement. "The basic *reasoning* by Rushton is solid evolutionary reasoning; that is, it is logically sound. If he had seen some apparent geographic variation for a *non-human* species — a species of sparrow or sparrow hawk, for example — no one would have batted an eye." But, Wilson added emphatically, even though society should be able to "handle" most areas of sociobiological debate, "when it comes to [human] racial differences, especially in the inflamed situation in this country, special safeguards and conventions need to be developed."

The great Mexican writer Octavio Paz once said that in literature what is truly important is *poetry* not the *poets*, who are, in the end, only fallible human beings like all the rest of us. Perhaps, in the Rushton affair, the same relationship holds true for *science* and its *scientists.* If so, it may be wise to address our most pressing questions not just to Rushton but to science and to all scientists. What are the responsibilities of scientists in a society that grants, as most of us would have it, the freedom of academic inquiry? What safeguards ought to be put in place, both in science and in society, to limit the obviously harmful impacts of certain kinds of ideas (the promulgation of bold theories of innate racial or gender superiority, for instance, or home recipes for cooking up lethal biological weapons)? ("Academics," says David Suzuki, "can no longer simply say I am searching for the *truth*; let the chips fall as they may. One *has* to be aware that human beings are influenced by those ideas.") And in a more practical vein, how might we begin to move toward a consensus on some sort of "ethical ground rules" for public and scientific dialogue over contentious notions like Rushton's? Ones that might permit both freedom of inquiry and expression, yet leave all of us less weary, wounded, and divided than many Canadians seem to be over Rushton?

Suzuki's passionate stand against Rushton in their debate earned him harsh criticism from many scientists on the sidelines. In retrospect, he expresses a few regrets about certain rhetorical tactics he used. But he says he harbors none for exercising his responsibility as "a member of the academic community to stand up and speak out and censure him." "What concerns me," says Suzuki, "is that Rushton has in effect been legitimized by the failure of the scientific community to expose him as someone whose claims have no scientific basis. Of course we have to be free to explore ideas. But there is an *added* burden of responsibility when those ideas have such clear repercussions as his do."

E. O. Wilson, besides being a leading sociobiologist has in recent years surfaced as one of biology's most eloquent spokesmen on the tragic consequences of the loss of species diversity through deforestation of tropical ecosystems; he suggests a need for an ethos that is sensitive not only to nature's variability but to *Homo sapiens*'. "I don't know the answer," Wilson told me, "but I think that for the good of humanity — as well as for ecosystems — we need to begin to see the tremendous strength in diversity and learn to celebrate that more as an ethic."

To do so, we must also remain vigilantly on guard against notions that though fully sanctioned by "objective" scientists, happen to conveniently grant one group unacceptable levels of power over others. Many people prefer not be reminded that some of the "scientific" foundations of Nazi Germany's genocidal policies for maintaining racial purity had been laid down, in part, by some of the most highly regarded academics of the day. Or that racist sterilization and anti-Semitic immigration laws in both Canada and the U.S. earlier in this century could easily have found similar rationale in the contemporary writings of some leading geneticists. Robert Jay Lifton, the Yale psychiatrist and author of several books on war and its psychological aftermath, in an article written in 1982 attempted to shed some light on exactly *how* scientifically sophisticated physicians in Germany had been able to rationalize their participation in state-sanctioned mass murder at concentration camps like Auschwitz. The answer, he found, lay in their deep ideological conviction that, as one Nazi SS doctor put it so paradoxically, they were killing "out of respect for human life." In their minds, of course, "life" did not refer to one individual but rather to the larger "life" of the nation as "cultivator of genes."

Racism, it has been said, is "softer," somehow less visible, in Canada than it is in the U.S. And that may be true in certain respects. But the record of racism in Canada, however quiet, is also unequivocal. Historically native peoples have been slaughtered and slandered here for centuries, often strictly on the basis of race, and beginning in the seventeenth century, there existed a viable trade in black slaves in these lands. During the 1920s the infamous American white supremacist group Ku Klux Klan began to find some Canadians quite receptive to membership, first in Anglo-Protestant Montreal, then spreading to many other parts of Canada, including areas not far from Rushton's own academic base in Southwestern Ontario. And if you have any doubts about the powerful impact of the hysterical wartime internment of Japanese Canadians in British Columbia in 1942, you might try asking Suzuki about the source of at least a portion of the deep sense of outrage he has exhibited toward Rushton's views on race.

But more useful than pointing out documented instances of racism in Canada, present or past, might be simply to encourage each individual — no matter how hopelessly romantic it might sound — to try to develop a greater sense of compassion and empathy toward all visibly different others. Once, in a moment of striking candor, Rushton, I thought, voiced a similar view in regard to his life as a scientist. As my tape recorder whirred, I told him that I could not help wondering whether he lacked any sense of the *pain* that his statements on race had inflicted on those most maligned by his hierarchical model. (I was thinking, as I spoke, of a scene during the Suzuki-Rushton debate when a local minister who happened to be black had angrily confronted Rushton with a tale of the anguish that his 11-year-old son had already experienced as a result of Rushton's statements on race. The young boy, an academically average student in a Grade 5 class that was otherwise white, had upon hearing about Rushton's views asked his father if they meant that he was destined to be "the dumbest person" in his class.)

Are you, I asked Rushton, as insensitive as some of your most severe critics suggest?

There might be an element of truth in such accusations, Rushton conceded. In fact there had been moments in the past when he had thought that he might have been better off as a scientist in some far-off "Shangri-La," where one could endlessly pursue any area of study one chose with no thought of the turmoil it might cause in faraway society. "*Shangri-La*," said Rushton softly. "It *does* appeal to me. I don't think I'm very socially oriented. I find it very difficult to form political judgments on things like abortion or the death penalty. I guess I *abstract* the issue."

I pressed on. Did he, as he had suggested in passing the day before, actually feel that there was "something missing" in his character as a scientist, that he was somehow lacking in empathy for some people?

"Yes!" he exclaimed. "I mean, I don't like to think of myself as an insensitive person. I don't think I am," he said, pausing to reflect on his words. "But I am *cerebral* — I think *that's* essentially it."

It has occurred to me more than once since our conversation that his statement is as much a measure of science — at least as it has been pursued by some overzealous members of the scientific community, especially when their concerns have focused on matters of race — as it is of Philippe Rushton, the man.

1989

Part V

ON LESSONS FROM NATURE

Introduction

THE VERY TITLE of this section, "On Lessons from Nature," is likely to arouse almost reflexively the ire of many members of the scientific community. I speak of those people, some of them distinguished scientists of the highest order, who steadfastly contend that science (by *its* very nature) can *only* see nature as utterly devoid of meaning and bereft of clues concerning human ethical conduct.

Neutral. Without any kind of message, however subtle, for *Homo sapiens*. Morally mute.

And any notion that anyone in the world might profitably seek *wisdom* in nature is generally dismissed with little more than an impatient wave of the hand and some vague reference to "naturalistic fallacy" — the nineteenth-century logical "proof," with widespread currency among scientists, that such matters have been rationally set to rest long ago, once and for all.

Well, if the scientific worldview *must* assign questions of human values and questions of science to strictly partitioned realms of human knowing, so be it. But that dogmatic Western perspective, however fruitful to the working scientist at his laboratory bench, need not, of course, shackle all the rest of humankind as we ponder pressing ethical problems in areas such as medicine, genetics, and the environment. That is, contemporary Western views regarding nature need not be the *only* ones around the world that happen to be reconcilable with the modern machinery of science and technology. Nor are they necessarily the *only* ones that could conceivably help humankind navigate the maze of science-generated ethical dilemmas that daily confront us.

While such possibilities are commonsensical to many peoples in the world, they can sometimes prove especially troubling to the so-called "objectivity" of the scientist. However, I think it is fair to suggest that growing numbers of scientists have begun to concede that there is room within the scientific worldview to look to nature for at least an "indirect" form of what might be called "moral guidance." How? At the very least, I would suggest, at the level of the *metaphor* — that literary device, quite legitimate even among the most rabidly ethnocentric of Westerners, that can be defined as an imaginative, insightful leap of analogy that can surface in the mind of any perceptive nature-watcher, regardless of his or her skin color, religious views, or geographic locale. The poetic imagination, in short, might be one science-compatible path to nature's inherent "wisdom."

That said, I simply offer the following two pieces as examples of my own clumsy search for meaning in the structure and processes of nature, however "logically fallacious" the quest might seem to some. The items, it turns out, seek "lessons" regarding variations on a single theme: the question of human responsibility for the fundamental units of biological heredity — genes.

The first — titled "Redesigning Life" — searches for lessons in the course of reviewing a range of issues arising from emerging genetic technologies from gene therapy and the designing of novel life forms to sequencing the human genome. And the second — "Blaming Crime on Chromosomes: The Mystery of the Man with Too Many Y's" (one of my favorites) — tries to distill a few humble lessons (that blasphemous word again!) in the compelling case history of the clash between new genetic techniques and human values in the so-called "XYY controversy" of recent decades.

Redesigning Life

HIGH ON A GRASSY KNOLL in rural Beltsville, Maryland, surrounded by acres of emerald pastures, rustling oak groves and herds of peacefully grazing cattle and sheep, stands a cluster of pre–Second World War vintage buildings housing the U.S. Department of Agriculture's Animal Science Institute. A leisurely half-hour drive from the headline-grabbing political turmoil of downtown Washington, D.C., it seems an unlikely locus of controversy. But beneath the surface of the pastoral scene lies a simmering conflict that could eventually erupt into a volatile, and global, public debate. The issue: the disturbing ethical, environmental, and legal questions arising from the rush to find profitable applications of revolutionary new genetic-engineering techniques — the most powerful and precise biological tools ever devised for redesigning life.

Inside a large red-brick barn bearing a sign that reads Quarantine Area/No Visitors and down a cool, foul-smelling concrete corridor, there are a number of third-generation offspring of, in the words of my soft-spoken, bespectacled host, Vernon Pursel, a physiologist, the "first successful gene-transfer experiment in the world using farm animals." Dozing fitfully and grunting as we approach are normal-looking animals that harbor not only the usual quota of porcine genes found in most pigs but also copies of an artificially inserted gene that normally stimulates growth in the human body.

The transgenic pigs are the product of a pioneering experiment, immortalized on the cover of the June 1985 issue of *Nature*, the prestigious British science journal. Scientists from the University of Washington, the University of Pennsylvania and the U.S. Department of Agriculture had successfully injected duplicates of the human growth-hormone gene directly into the nuclei of living fertilized pig eggs, or zygotes, using a thin glass needle. Spliced to the human gene was a regulator region of mouse DNA designed to activate the gene in the liver, intestines and other organs, rather than in the pituitary gland, where the hormone is normally secreted. The altered eggs were then transferred into the oviducts of surrogate mothers — hormonally stimulated female pigs — on the Beltsville farm.

Only about one in 10 of the developing embryos successfully incorporated the human gene into their chromosomes. But in roughly 60 percent of the transgenic piglets, the foreign gene functioned correctly, resulting in measurable levels of human growth hormone circulating in their blood.

A major goal of the project had been to create genetically modified pigs that, boosted by the addition of human growth-hormone genes, would grow faster and more efficiently and would reach a substantially larger body size. Only a few years earlier the most prominent members of the Beltsville team, Richard Palmiter, a molecular biologist, and Ralph Brinster, a reproductive physiologist, had carried out experiments in which copies of the human growth-hormone gene were injected into fertilized eggs of laboratory mice. The results had been stunning. Some of the genetically engineered "supermice" harboring the active human gene had growth-hormone levels as much as 1000 times higher than normal and grew up to twice the average size.

The Beltsville pigs also had increased levels of human growth hormone in their blood, both before and after birth. But they did not display the expected surges in growth; the genetically altered swine were leaner but no larger than normal control animals. In fact, the experimental pigs displayed several distinct defects, apparently resulting from a surplus of the human hormone spewing from the poorly expressed gene. They were noticeably less active than their normal counterparts. A disproportionate number of them developed ulcers and seemed unusually susceptible to infection and arthritis. In the case of females, some proved to be infertile.

"We simply haven't seen enhanced growth rates in these transgenic pigs," says Pursel. "Instead, we see abnormalities in early bone growth in that the leg bones are actually shorter and larger in diameter. In addition, these transgenic animals have a depressed appetite and are very definitely lethargic. They spend most of their time lying down. You might say that these pigs are behind the eight ball from the moment of birth."

To most of the growing legions of scientists around the world engaged in gene-splicing research, Beltsville's transgenic pigs were intriguing but hardly earthshaking. They represented little more than the logical extension of a wave of similar experiments aimed at transporting foreign genes into complex multicellular animals and plants — from fruit flies and frogs to domestic food crops. Yet not long ago, such feats of genetic engineering would have seemed the stuff of science fiction. Consider just a few dramatic examples:

- In 1987, scientists at the American National Institutes of Health succeeded in inserting copies of a human gene into monkeys, where the gene directed the synthesis of a human enzyme. In the same year, newspaper reports appeared in the United States of clandestine attempts by scientists to create hybrid human-ape embryos by artificially fertilizing chimpanzee ova with human sperm. The claims have remained unsubstantiated. But they continue to reflect deep public concern over the possibility that researchers may one day be tempted to concoct, in the words of a respected U.S. presidential commission report on ethical issues in genetic engineering, highly intelligent but servile interspecies hybrids, or genetic chimeras: "a group of virtual slaves — partly human, partly lower animals — to do people's bidding."
- In laboratories in Canada, the United States, Great Britain and elsewhere, scientists are devising techniques for transforming cows and other large mammals into living pharmaceutical factories. The idea is to equip the animals with human genes coding for medically useful substances such as rare blood-clotting factors, insulin, or growth hormone — but ones that have been chemically modified so they direct protein synthesis only in specific host tissues. If the foreign genes were expressed only in

a cow's mammary glands, for example, the finished human protein might then be harvested in commercial quantities simply by milking the cow and extracting the medicine.

- In the past four years scientists at the University of California, Davis, have created nine novel Alice in Wonderland creatures they call "geeps" — ungainly, goat-faced, woolly-bodied chimeras made by fusing bits of a goat and a sheep embryo in a test tube, then implanting the composite embryo in a goat's womb. Depending on whether female geeps are bred to a sheep or a goat, they will produce either a lamb or a kid.
- Scientists at Canada's Plant Biotechnology Institute, a national research facility in Saskatoon, Saskatchewan, and at other laboratories around the world are fine-tuning techniques for using bacteria to ferry foreign genes directly into higher plants. Among the genes under study are ones coding for insecticides, plant resistance to commercial herbicides and other useful traits that could enhance the commercial value of certain domestic crops.

Human beings have, of course, enjoyed a long tradition in biotechnology in products ranging from industrial chemicals and yogurt to beer and antibiotics. And hereditary traits in animals and plants have been shaped by artificial selection for millennia. Until very recently, however, efforts to redesign organisms to meet human needs have been based on relatively crude attempts to control the transmission of characteristics within a species or between evolutionarily closely related species.

The ability of modern molecular biologists to cut, edit, splice, and move genes between virtually any species on Earth — even the most distantly related — represents a quantum leap in humankind's ancient tradition of genetic control. To a biologist engaged in basic research in genetics, the calculated exchange of tiny pieces of genetic material between different species represents a powerful new tool for isolating the activity of a single gene within the constellation of tens of thousands of genes that interact during the growth and development of a higher plant or animal. It also offers the possibility of producing genetically altered laboratory animals designed to mimic certain human genetic diseases that are otherwise difficult to study.

Public reaction to such scientific prowess has not been uniformly enthusiastic, however. The Beltsville scientists, for example, were unable to visually distinguish transgenic piglets from normal control animals and had to rely instead on sophisticated laboratory tests to confirm the presence or absence of the human gene. Yet after accounts describing the experiments appeared in *The New York Times* and *Newsweek* — in terms the scientists viewed as somewhat sensational — the response sometimes bordered on horror.

"People started saying, 'Oh, your pigs have human genes!' " sighs a beleaguered Pursel. " 'Do they also have human ears? Do they talk?' Some people even phoned us and announced angrily that they would never buy pork in the supermarket again. Of course our transgenic pigs were never intended to be marketed. But the very idea of an animal carrying a human gene was something entirely new to them."

UNDERSTANDABLY SO. After all, it was only 14 years ago that Stanley Cohen of Stanford University and Herbert Boyer of the University of California, San Francisco, first snipped a piece of DNA from a frog and spliced it into the genetic molecules of a bacterium, where the foreign gene functioned. By fusing DNA molecules from two different organisms and species, they not only had created a hybrid DNA molecule known as recombinant DNA but also had engineered a novel life form that was unknown in nature. Moreover, by demonstrating that the genes and proteins of virtually any species could be mass produced, or cloned, in living bacteria cells, the experiment launched a new age in biotechnology based on the manipulation not simply of biological processes but of the hereditary molecules that orchestrate them. The implications of the new age proliferated last year when the U.S. Patent and Trademark Office decreed that man-made animals could be patented.

Society has already begun to reap the rewards of recombinant DNA and related genetic technologies. In 1982, human insulin, the world's first gene-cloned medical product, was approved in the United States for general use, replacing the drug required by diabetics that had previously been painstakingly derived from the pancreases of cows and pigs. And since then a burgeoning global biotechnology industry has developed and manufactured, in rapid succession, a host of other genetically engineered products — ranging from new vaccines and prenatal tests for certain human hereditary diseases to a growing pharmacopoeia of cloned natural medicines. Among them: human growth hormones to treat inherited growth deficiencies, interferons to combat certain forms of cancer, and, most recently, potent blood-clot-dissolving substances useful in treating victims of heart attacks.

As the exploits of gene splicers become more daring and as the potential implications of their research become clearer, governments around the world are scrambling to update existing laws and regulations to provide safeguards that better reflect the current state of the art. One of the main cases put forward by those who advocate tougher controls over, for example, the release of man-made organisms is that — much like exotic agricultural weeds, insect pests or microbes from faraway places — the introduction of novel, self-perpetuating life forms could conceivably disturb local ecosystems in unexpected ways.

Genetically altered livestock, of course, would seem unlikely candidates for persistent ecological pests. Their movements would be severely restricted by barns and fences, and domestic animals would generally be relatively ill equipped for an independent existence if they escaped confinement. Even in the most farfetched scenario imaginable — herds of giant transgenic swine thundering across rural landscapes — the only environmental clean-up necessary might be carried out with no more than rifles and higher fences.

Worst-case environmental scenarios would immediately become more sobering if a foreign gene were transferred into a wild, rather than a domestic, species. "In principle," muses Palmiter, the molecular biologist, as he tilts back in his chair in his office at the University of Washington, "you could apply the same sort of gene-transfer strategies to wild-animal species. But I don't think there is much economic incentive to do so. I imagine the first place you are likely to see that would be with researchers trying to make transgenic fish. A transgenic fish would be a lot easier to lose track of than a transgenic cow."

To date, however, most of the controversy generated over

possible long-term ecological impacts of releasing genetically engineered organisms has not centered on novel vertebrate animals. That is still a relatively rare genetic application. Rather, the debate has focused on genetically altered microorganisms and, to a lesser extent, on genetically altered forms of domestic plants. In 1987 alone, a number of reports of attempts by scientists to conduct open-air field tests of genetically modified bacteria caused an international furor in some circles.

One of the most widely reported cases concerned efforts in the United States in 1987 to test a genetically altered variety of bacterium designed to reduce frost damage in potato crops. Known as Ice-Minus, the microbe — a member of a common bacterial species called *Pseudomonas syringae* — was first developed by Steven Lindow, a plant pathologist at the University of California, Berkeley, who discovered that by removing a single gene he could transform the organism from a frost inducer to a frost inhibitor. Amid considerable controversy, potato seeds treated with the bacteria were planted at a remote field station near the Oregon border on April 29, 1987. Less than a week earlier, a similar experiment had been conducted in a California strawberry plot.

An unauthorized field test also made headlines in North America last year after Gary Strobel, a plant pathologist at Montana State University, announced that — as an expression of "civil disobedience" against federal Environmental Protection Agency (EPA) regulations he considered too restrictive — he had injected genetically altered bacteria into 14 young elm trees growing in a nursery behind the campus football stadium. The novel microbes were designed as a biological control for Dutch elm disease. But by September, in the face of public outcry, Strobel, in tears, voluntarily cut down and incinerated his stand of experimental elms to avoid damaging his university's reputation. He was later reprimanded by the EPA.

In Canada, there are no laws specifically covering the deliberate release of genetically engineered organisms, a situation that astonishes Andrew Kimbrell, policy director of the Washington-based Foundation on Economic Trends, an organization that has spearheaded public opposition in the United States to a wide range of gene-splicing projects. "I think it is very surprising," he says. "After all, this whole conflict over genetic technologies does not exist in a historical vacuum. In our lifetimes we have witnessed revolutions in nuclear and petrochemical technologies, and in each case the hard questions were not asked prior to the commercialization of those technologies. Now we have a new technology with extraordinary potential risks, and we seem unwilling to ask the hard questions. We must realize that technology itself can be a form of legislation; it legislates our lives as surely as political legislation does."

While the Canadian government appears reluctant to adopt regulations that might hinder the development of the country's fledgling biotechnology industry or upset delicate free trade deals, it has recently begun to review existing statutes and guidelines applicable to biotechnology — ones administered by Agriculture Canada, Environment Canada, Health and Welfare Canada, and the provinces.

One purpose of the review process, according to David Shindler, manager of biotechnology for the Ministry of State for Science and Technology, is to prepare what he refers to as a "regulatory road map" to guide researchers through the process of government notification and review. At the same time, the review is attempting to spot gaps in existing legislation — much of it drafted decades ago to provide safeguards against inanimate chemical pollutants that although they might linger in the environment are incapable of self-replication like a novel life form.

In the meantime there is little indication that biotechnology companies are taking advantage of Canada's current *laissez-faire* attitude toward field testing of genetically altered organisms. There are, though, a few isolated rumors: a spokesperson for one leading Canadian biotechnology firm told me almost in passing that he knew of at least two separate field tests of genetically altered plants in Canada during the summer of 1987. The tests, he said, were carried out by two large American corporations with the knowledge of Agriculture Canada officials. But spokespeople from both firms denied that any novel organisms have been released in this country. So, too, did several key officials involved with Canadian federal supervision of biotechnology. "It has not happened," said Shindler. "There have not been field tests in Canada."

Shindler's position is echoed by Martin Boddington, head of Environment Canada's Biotechnology Centre and chairman of a federal working group established in 1985 to assess Canadian safety and regulatory issues related to biotechnology: "While there may have been a case like Strobel's in Canada, we don't know about it." Environment Canada's proposal for a new Canadian Environmental Protection Act is intended to control the environmental release of any novel substance — "whether animate or inanimate." Until the act is enacted into law, possibly as early as spring 1989, says Boddington, the agency will continue in an atmosphere of legal uncertainty.

Canadian law also lacks provisions to allow a scientist or company to claim ownership of a novel multicellular animal or plant. The shift last year in the U.S. patent policy has the potential to influence similar legislation here. The landmark American decision, contained in a brief memorandum signed in early April, proclaimed that for the first time, the U.S. patent office was prepared to examine claims for granting patent rights for evolutionarily more advanced "multicellular living organisms, including animals," that were "a product of human ingenuity." The genetically altered forms of only a single species — *Homo sapiens* — were explicitly excluded from the new U.S. patent policy on the grounds that ownership of human beings is "prohibited by the [U.S.] Constitution."

If similar patent policies were eventually adopted worldwide, the economic impact of the trendsetting U.S. stand could be enormous for Canada and other nations nurturing young biotechnology industries. It might provide some private biotechnology firms with a chance to recoup the funds they are investing in gene-splicing research on the genetic manipulation of higher animals and plants, and it would certainly provide added incentive for profits. "The Americans recognize something we Canadians don't — that patents are an economic tool," says Francis Rolleston of the Medical Research Council (MRC). "The point of a patent is to confer economic advantage." Critics counter that it might alter the face of agriculture around the world for years to come — compelling farmers to rely on an ever smaller number of lines of "superior" livestock available only on payment of royalty fees to a corporate patent holder. And by granting developed countries a chance to consolidate their technological lead over less developed countries, it could add to the already paralyzing dependency of the poorest nations on richer, more scientifically sophisticated trading partners.

To some observers, including the longtime American activist Jeremy Rifkin, the notion of granting permission to patent a living animal evokes ethical concerns that loom every bit as large as economic ones. Do human beings, they ask, have the right to own the genetic makeup of an entire potentially immortal lineage of animals? And does the mere addition or alteration of a single gene or gene fragment somewhere in the tens of thousands of genes making up a multicellular animal's genome — the legacy of millions of years of that species' evolutionary history — transform an organism into a product of "human ingenuity?"

"If a researcher implants a foreign gene into an animal's genetic code — for example, a human growth hormone," observes Rifkin, "the genetically altered animal is considered a human invention, like a toaster, automobile, or tennis ball." Adds Kimbrell: "There is in the very notion of humans' 'manufacturing' a genetically novel organism an element of what I like to call inanimism. You know, we use the term animism for people who ascribe the qualities of living organisms to nonliving things. But we haven't yet coined a phrase for the reverse — the confusion of living things with inanimate matter.

"From our point of view," continues Kimbrell, "the U.S. government has now created a powerful incentive for private companies to rush out and create genetic chimeras. Yet we don't have adequate regulations in place. As a result, we are going to have our regulatory agencies frantically chasing this technological problem. We may have managed to get away with doing this in the past, but it is certainly no way to run a technology — or an environment."

IN THE SHADOWS of the debate over issues surrounding genetic manipulation of multicellular organisms lie deep-seated fears that emerging genetic-engineering techniques will eventually find application in the modification of human beings. Considering the relative ease with which certain reproductive technologies — from the frozen storage of sperm to in-vitro fertilization and embryo transfer — have made the transition from domestic animal subjects to human patients, not to mention humankind's historic fascination with the notion of genetically "improving" *Homo sapiens*, such concerns should not be casually dismissed. In principle, variations of many of the techniques available to researchers transferring foreign genes into laboratory mice or livestock could also be used for human gene therapy — to remove, replace, or repair a defective human DNA sequence involving one of the 3000 or more inherited diseases medicine is currently able to detect. But for now, at least, only gene-transfer techniques that would permit medical manipulation of human body, or somatic, cells — in contrast to reproductive, or germ, cells capable of transmitting a new trait to future generations — are under active consideration.

The prospect of human gene therapy has been greatly enhanced in recent years by the remarkable progress geneticists are making in deciphering the exact sequence of the subunits that make up the estimated 100,000 genes in a typical human body cell and in determining the location of disease-related genes. By deciphering the DNA code, geneticists have been able in some cases to pinpoint the precise location of hereditary defects. In others they have succeeded in at least identifying neighboring DNA sequences that serve as reliable early indicators, or markers, for hereditary disease. And by isolating and cloning individual human genes they have begun to build up a repository of healthy "replacement" genes for future genetic repairs.

Techniques for charting the entire system of human genes, or our genome, are progressing so rapidly that many experts are suggesting by the year 2000 science could possess a complete map of the three billion or so bases that constitute the total genetic content of our species. That genetic information can be reduced to nothing more than a long computer printout of row after row of the various combinations of the four possible nucleotide bases. Such a coordinated all-out initiative to decipher human DNA could have a price tag of more than $2 billion (U.S.), require huge supercomputer data bases for information storage, and represent a megaproject in the same league of big science as sending an astronaut to the moon.

Not surprisingly, as private companies have entered the global race to crack the human genetic code, some disturbing legal questions reminiscent of those surfacing in the ongoing debate over patenting animals have arisen. Ultimately who — if anyone —

owns the human genome? In 1987, Walter Gilbert, an internationally respected Harvard biologist and a pioneer in DNA sequencing techniques, publicly announced that his new company, Genome Corporation, intends to seek copyrights for its human DNA sequencing data. Genome Corporation, said Gilbert, to the dismay of some scientific colleagues, "will create a catalog of all human genes [which] would be made available to everyone for a price."

Less than a decade ago, the notion of treating human diseases by performing what could be called genetic surgery might have seemed absurdly futuristic. But there is now a growing consensus among medical geneticists that clinical trials involving direct manipulation of human DNA are only a matter of a few years, perhaps even months, away. W. French Anderson, a leading researcher in the field based at the National Institutes of Health in Bethesda, Maryland, and an outspoken proponent of early clinical trials, says that initially the most likely targets for gene therapy are a number of fatal hereditary illnesses affecting blood-forming cells of the bone marrow — one of the few body tissues that can, at present, be removed, treated, and replaced.

One of the leading candidates is severe combined immunodeficiency disease (SCID), an inherited condition that results in drastically lowered resistance to infection in young children — so-called bubble babies who must be reared in germ-free environments. Victims of SCID harbor a defective gene for the enzyme adenosine deaminase (ADA), a product of bone-marrow cells, which is an essential ingredient in a healthy immune response. If untreated, children usually die from infections within the first two years of life. It is possible to correct the debilitating disease with intravenous injections of normal ADA-synthesizing bone-marrow cells from a matched donor. But the goal of the gene therapist would be to transplant genes, rather than tissue.

A number of Canadian scientists are also poised to perform human gene therapy. They include Alan Bernstein, at Mount Sinai Hospital Research Institute in Toronto, and Keith Humphries, at the Terry Fox Laboratory of the British Columbia Cancer Research Centre in Vancouver, along with their colleagues. "Canadian scientists are no slouches," says Rolleston of the Medical Research Council. "They are every bit as good as the Americans.

They foresee getting into the area of human gene therapy and have been asking what the MRC thinks about it."

The MRC has not yet formulated a comprehensive set of guidelines for prospective Canadian gene therapists, although the topic is raised briefly in existing guidelines for medical ethics and human experimentation. But the agency is establishing a special working group of medical, legal, and other experts who will explore critical ethical issues surrounding human gene therapy. It will be chaired by Patricia Baird, head of the medical genetics department at the University of British Columbia medical school, and may report its findings as early as this year. According to Baird, the pace of progress in medical genetics makes the prospect of human gene therapy almost inevitable. "Sometime in the not too distant future," she observes, "people will consider it appropriate to offer certain parents some form of gene therapy. So the MRC's proposal to anticipate these events by preparing a set of guidelines to advise researchers is very timely."

ORDINARY PEOPLE living lives far removed from the esoteric world of molecular biologists may accept without hesitation the value of certain forms of human gene therapy, but they may well wonder why a scientist would even *want* to place human genes in a pig. As we stood in the parking lot outside the Beltsville research facility beneath a towering stand of oaks in the final moments before my departure, I could not refrain from asking a researcher who had participated in the transgenic pig project what he would tell such people.

He paused and then, answering slowly, as if such concerns weighed more heavily on him than he had suggested in our earlier technical discussions of transgenic pigs, made a remarkably candid disclosure. Although there were intriguing problems to be solved by inserting foreign genes into livestock, he said, there were really no compelling scientific reasons to insert human genes rather than genes from some other species into the Beltsville pigs. As a matter of fact the human genes were used largely as a matter of convenience. The agricultural significance of the Beltsville project might actually have been greater if the researchers had used pig growth-hormone genes. "But cloned pig growth-hormone genes have simply not been available to researchers. In fact, most of the genes of farm animals are just not available. The first genes from mammals that are available to researchers are usually either mouse genes or human genes. That's why we used a human gene." Thus the fact that transgenic pigs at Beltsville harbor human rather than porcine genes may turn out to be due less to some noble search for scientific knowledge than to a trait that has become the hallmark of the headlong, hotly competitive global race to find new applications of genetics — haste.

Today, there is a growing recognition among many scientists and nonscientists alike not only that we are living in the midst of a technological revolution in terms of our ability to manipulate DNA molecules but also that these powers are beginning to force us to face difficult questions that have not been faced by any other generation in history. To admit this is not to declare oneself a scientific heretic or to deny the potential benefits that genetic technologies are likely to provide. But it does suggest an urgent need to start searching for a new, broad-based public consensus on some of the most divisive issues concerning the applications of modern genetics. And there is every reason to believe that the process will require considerable tolerance, candor and — perhaps most important of all — time. Time to weigh possible social and economic impacts of new commercial applications of genetics. Time to render arcane scientific techniques understandable to the general public. And time to encourage thoughtful public dialogue concerning the technological paths that are best suited to our collectively chosen future — even if it means deliberately delaying certain applications of genetics that most clash with cherished human values.

Given that time, perhaps we will not become unduly fearful of the sight — to borrow an image from William Butler Yeats's poem "The Second Coming" — of an approaching "shape with lion body and the head of a man." For in this tumultuous age of discovery and application in genetics, we have an opportunity to determine that what is created is not just a vision of some unearthly life form conjured up in the laboratory of a geneticist but rather a true reflection of our species' emerging knowledge — of the biological basis of what makes us human and of our proper place in the natural world.

1988

Blaming Crime on Chromosomes: The Mystery of the Man with Too Many Y's

Genethic Principle
The vast majority of human hereditary differences are polygenic, or involve the interplay of many genes; therefore, it is a dangerous simplification to proclaim a causal relationship between human behaviors and so-called "defects" in human DNA.

Years later, Dante was dying in Ravenna, as unjustified and as lonely as any other man. In a dream, God declared to him the secret purpose of his life and work; Dante, in wonderment, knew at last who and what he was and blessed the bitterness of his life. Tradition relates that upon waking he felt that he had received and lost an infinite thing, something he would not be able to recuperate or even glimpse, for the machinery of the world is much too complex for the simplicity of men.

Jorge Luis Borges, in "Inferno I, 32"

In 1961, a tall, healthy, normal-looking middle-aged man — with a history, it seems, of barroom brawling, especially during his youth — underwent a series of medical tests to determine why he had recently fathered a genetically abnormal child. The tests did not identify the cause of the baby's illness. But they did inadvertently reveal something that astonished the man's doctors. Every somatic cell in his body had one too many chromosomes — 47 — and in a combination, or genotype, that had never before been reported in the literature of medicine.

In the shorthand of the medical geneticist, the man's genotype was 47,XYY. In addition to the usual 46 chromosomes, including 44 *autosomes* (22 pairs), or chromosomes that are not sex chromosomes, and the X and Y sex chromosomes normally found in a human male, he had an extra Y chromosome.

The report of the world's first recorded case of an XYY man was published that year in the British medical journal *Lancet*, making minor medical history. Today we know that the XYY genotype is a relatively rare chromosomal condition, occurring in about one in a thousand newborn male infants. And we know that it can be traced to a failure in the distribution of chromosomes, called *nondisjunction*, that takes place in paternal reproductive cells during meiotic metaphase.

During act II of meiosis duplicate pairs of chromosomes normally separate and drift toward opposite poles of the dividing cell (see Figure 1). Occasionally, though, sex chromosomes fail to part and become trapped together, like Siamese twins, in one of the daughter cells; they may be two duplicate X chromosomes or two duplicate Y chromosomes. If the tandem structure turns out to be two Y chromosomes, the cell will eventually form a maverick YY sperm. When this sperm later goes on to fertilize an X-bearing egg from a normal female, the result will be a male embryo that has an XYY genotype. As the genetically abnormal embryo matures in its mother's uterus, it transmits its microscopic chromosomal error through mitosis to millions of other body cells.

The discovery of the XYY man was made possible by karyotyping, a revolutionary new laboratory technique developed during the 1950s that provided a direct microscopic inventory of human chromosomes. Before karyotyping, geneticists lacked any reliable method of detecting even gross chromosomal abnormalities such as the XYY condition. In fact, they did not even know for certain precisely how many chromosomes human body cells harbored.

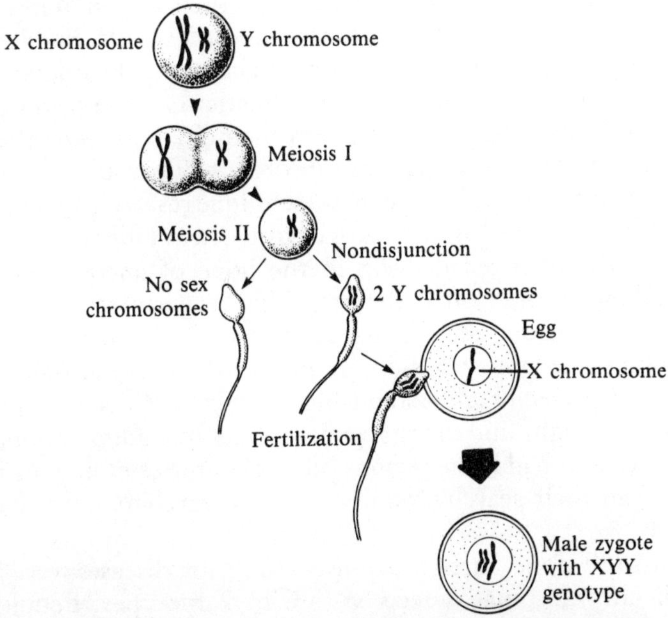

Figure 1: Meiotic nondisjunction leading to XYY male

In 1956, two researchers, Joe Hin Tijo and Albert Levan, made microscopic smears of lung cells from aborted fetuses that established with certainty for the first time that the human diploid number of chromosomes was 46. Their technique was quickly refined and adapted into an effective technique for assessing the chromosomes of adult patients.

The procedure was relatively simple. Human cells from laboratory tissue cultures or freshly drawn blood were treated with chemicals to halt the movement of chromosomes during midmitosis. Individual cells were then squashed, smeared across a glass microscope slide and stained for visibility. The entire quota of chromosomes in one cell could then be photographed through a microscope. The final step was to clip each chromosomal image from the print, sort the images according to size into rows of matched homologous pairs and paste them into a composite photograph known as a karyotype. Under such scrutiny, each pair of chromosomes in the human genome began to emerge with its own distinctive identity — based on its size, the location of its centromere, horizontal banding patterns and other traits. Suddenly, by comparing karyotypes of patients with inherited disorders to karyotypes of normal, healthy people, geneticists could begin to look for subtle structural differences in the human genome that they might correlate with a patient's disorder or disease.

Armed with this powerful new technique research teams quickly set out to produce karyotypes of victims of a number of perplexing, seemingly inherited diseases in the hope of uncovering corresponding chromosomal defects. They eagerly set out on this search — without a clear grasp of any causal relationships between chromosome abnormalities and human disease. But their eagerness was perfectly understandable, for it reflected a genuine hope that they might find the first genetic clues to a number of baffling diseases that had been responsible for centuries of human suffering. And their search would win the researchers immediate rewards.

In 1959 alone, three previously enigmatic diseases were linked with chromosomal defects visible in karyotypes. People with Down's syndrome, a congenital condition that could leave children mentally retarded and weakened by a constellation of physical problems, were found to possess an extra autosome. They all had a triplet set, or *trisomy*, of chromosome pair number 21; other chromosomes were normal. Patients with Klinefelter's syndrome, a rare disease that tends to feminize breast development and other secondary sex characteristics in men and render them sterile, turned out to have an XXY genotype. And women afflicted with Turner's syndrome, a condition that causes infertility and immature sexual development, proved to have XO genotypes, lacking the second X chromosome typical of females. All these conditions would eventually be traced, like the XYY disorder, to errors in the orderly partitioning of chromosomal pairs during meiosis (see Figure 2).

These were intoxicating discoveries for medical geneticists and left them justly proud. And these discoveries were quickly followed by others. In this atmosphere of optimism it is not difficult to imagine the flurry of excitement created by the brief clinical report in *Lancet*, just two years later, of an anonymous, slightly aggressive middle-aged man who had been born with a previously unknown

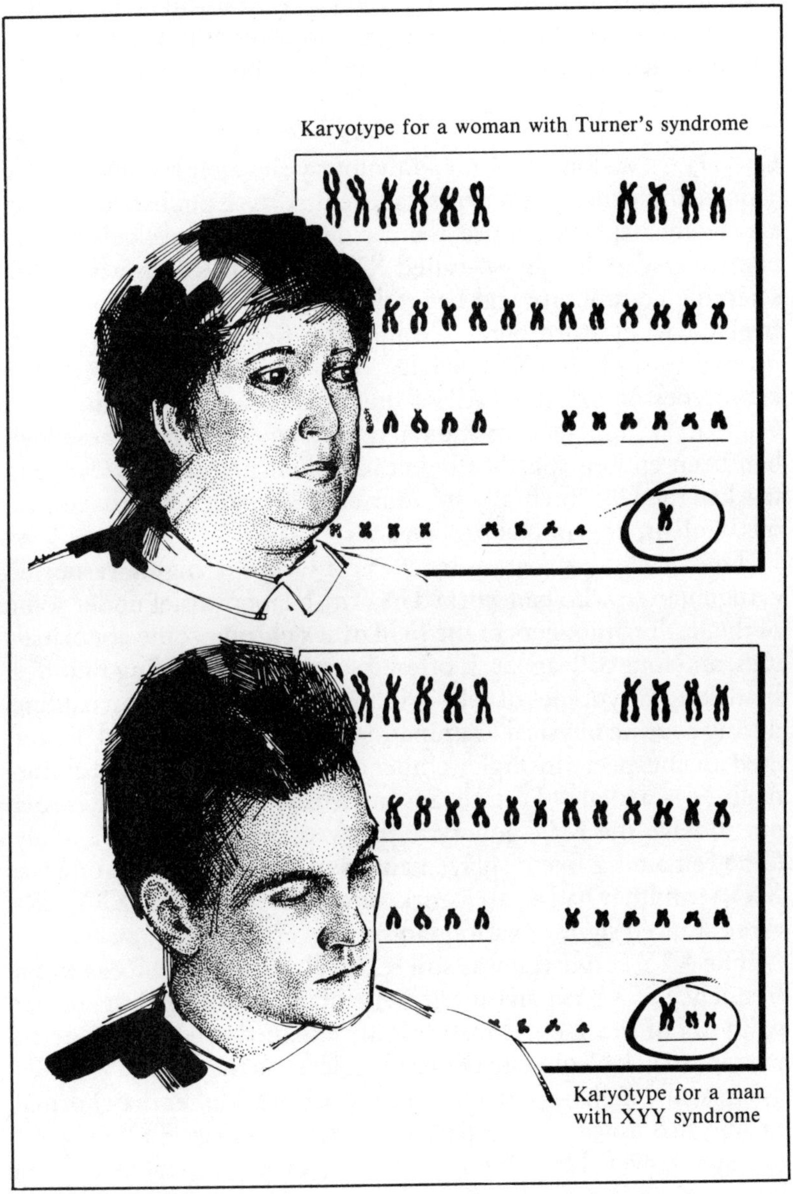

Figure 2: Karotypes of XXY and Turner's syndromes

sex chromosome defect known as XYY. In the same journal, less than a year later, the tantalizing proposition was put forward that the same genotype might possibly be linked to a life of crime.

ON DECEMBER 25, 1965, only four years after the initial XYY report, the prestigious British journal *Nature* published a short article that sparked more than a decade of fierce debate and controversy about the so-called XYY syndrome both within the scientific community and beyond. Titled — somewhat provocatively, it would turn out — "Aggressive Behaviour, Mental Subnormality and the XYY Male," it described a survey of the karyotypes of 197 men. All of them happened to be patients in Carstairs, a high-security mental hospital in Scotland. These men had been chosen specifically because their presence at Carstairs marked them as "mentally sub-normal male patients with dangerous, violent, or criminal propensities."

The article was written by Patricia Jacobs, a highly respected cytogeneticist who had worked in British laboratories under some of the leading pioneers in the field of sex chromosome abnormalities, and four colleagues. It offered some rather startling findings. In making karyotypes of cells in blood samples taken from patients during routine physical examinations, the researchers had discovered an unexpectedly high number of visible chromosomal abnormalities — a total of 12 in all. Seven men, about 3.5 percent, turned out to have the XYY genotype that was the target of the study. (The remaining five displayed an assortment of defects: one was XXYY; another had a patchwork, or mosaic, of XY and XXY cells; three showed signs of various autosomal errors.)

The XYY karyotype was still relatively new to science, and the frequency of XYY males in the general public, beyond the barred windows of the mental institution, remained unknown. But 3.5 percent struck Jacobs and her team, at least on theoretical grounds, as unusually high — perhaps 20 or more times higher than normal. Jacobs also noted almost parenthetically that her XYY subjects, averaging over 1.8 meters, or 6 feet, in height, seemed to be noticeably taller than genetically normal males. But without a doubt the most damning factor in this XYY study was the uniform history of mental illness, aggression, or both in its subjects. That history was officially documented; now so was the high incidence of XYY karyotypes. While the relationship between the two might remain uncertain, the implications seemed less so. "At present," concluded the authors, "it is not yet clear whether the increased frequency of XYY males found in this institution is related to aggressive behavior or to their mental deficiency or to a combination of these factors."

Despite the note of uncertainty in the statement, it nonetheless seemed to imply that the XYY genotype had already been linked to male aggression or mental defects. Earlier studies of related sex chromosome defects had already speculated on the possibility of such a relationship. But now, for the first time, the association between XYY males and abnormal behavior was indelibly recorded in the pages of one of the world's most highly regarded scientific journals. Could one extra Y chromosome somehow alter the genetic destiny of a newborn male child — predisposing him to a life clouded by mental illness, violence, or crime? And if so, why? Moreover, what ought responsible, genetically "normal" citizens do to protect their communities from the possible dangers of these free-roaming hordes of "chromosomal defectives"?

WHAT WAS DISTINCTIVE about the XYY genotype from the beginning was that unlike more severe conditions such as Down's syndrome and other serious chromosomal defects, it was accompanied by no predictable set of physical or mental symptoms that could be used to reliably diagnose the condition. Down's syndrome, for instance, was not detected solely by the presence of an extra chromosome 21. People with Down's syndrome might display any of a number of defects from distinctively slanted eyes and a deformed skull to mental deficiencies and congenital heart disorders. In contrast XYY males would turn out to display few predictable characteristics. The most notable: increased height and lowered fertility when compared with genetically normal males. Most other stereotypical XYY symptoms would later be traced to anecdotal evidence, conjecture, or statistical sleight of hand.

The first documented XYY case had not been diagnosed on the basis of bizarre physical or behavioral symptoms but inadvertently by medical fluke. Nor did Jacobs's influential but preliminary study pretend to establish a firm causal link between the XYY genotype and the tragic behavioral histories of the men at Carstairs. No one had yet begun to dissect the complex question of what, if anything, genes had to do with those behaviors.

Part of the excitement over the XYY syndrome lay in the hope that it might eventually shed light on broader questions about the genetic basis of behavior in normal XY males. The line of reasoning went something like this: men have a Y chromosome that women lack. Men tend to be more aggressive than women. If the Y chromosome were instrumental in altering sex-related bodily characteristics — from beard and biceps to the male sex organ — perhaps it was also essential for some of the behavioral patterns we consider masculine. And if it were, perhaps the newly discovered XYY male — in a sense, a "supermale," with his double dose of the masculinizing Y chromosome — would serve as a natural test case to examine that hypothesis.

The logic of this hypothesis, it should be noted, highlights the importance of genetic factors in gender differences in human behaviors. There was, and continues to be, intense debate among respected biologists over the relative contribution of heredity and environment to the sex roles that males and females eventually assume in society.

We now know that most of the Y chromosome consists of *heterochromatin* — condensed regions of DNA containing genes that are inactivated and metabolically silent. Other stretches of DNA are complementary to those carried by its homologous partner, the X chromosome. Among the remaining relatively small number of genes unique to the Y chromosome are those that determine genetic maleness. During embryonic development, these genes activate the development of male reproductive organs, the production of sperm and the secretion of the masculinizing hormone testosterone. So the question might be refined to this: could behavioral studies of XYY males, who presumably had twice the usual number of Y chromosome genes that confer genetic maleness, also illuminate the role of those genes in the behavior of normal males?

It was not a frivolous question. Earlier karyotype surveys of mentally retarded men in Swedish and British mental institutions had already alerted Jacobs to the possibility that an extra Y chromosome might predispose a man to aggressive behavior. In these studies of men with an extra X chromosome, those who also had a surplus Y chromosome (for example, XXYY genotype) were more likely to be found in high-security than in ordinary mental hospitals, compared with men having one Y chromosome (for example, XXY genotype). This finding had seemed to Jacobs to pose a question of almost mathematical clarity: did one extra Y transform the multi-X genotype into a formula for aggression?

Nor is there any reason to believe that Jacobs's hypothesis was in any sense mean-spirited. Her stated aim was simply to pursue this line of reasoning to its logical conclusion. If the extra Y did somehow contribute to a more aggressive male personality in multi-X genotypes, the effect might be clearer in the simpler XYY genotype. Yet, whether well-meaning scientists were aware of it or not, the seeds for a bitter debate over XYY research were sown by the very nature of the Jacobs study.

First, the study focused on forms of human behavior — male violence and criminality — that represent a potential threat to society. Clearly if karyotyping could somehow be used to quickly screen a population for early genetic indicators of antisocial behavior, it would be of enormous interest to police and other security forces. Such a program could also become a dangerous precedent leading to increased efforts to link other socially unacceptable behaviors — political terrorism, alcoholism, drug abuse, even homosexuality — however tenuously, to genetic markers that could be screened.

Second, the Jacobs study set out to establish a genetic basis for forms of behavior considered negative by most people. From the beginning the XYY genotype was associated with disturbed, violent men in high-security insane asylums. Anyone found to possess the genotype tended to be stigmatized as violent or mentally ill — whether it was true or not. If the XYY abnormality had been investigated for a suspected association with more praiseworthy human behaviors — say, entrepreneurial talent, leadership skill or athletic prowess — any hasty hypothetical association between chromosome and character would have been less likely to cause personal harm. But once a young boy, for instance, was found to possess an XYY genotype, he could be marked for life — burdened with an unearned scarlet letter in the form of a second, or supernumerary, Y.

Finally, the XYY study proposed to apply what at the time was one of the most powerful new tools in medical genetics — karyotyping — to one of the most politically powerless groups in our society — prisoners who were mentally ill. It could be argued that they were precisely the people who eventually might benefit most from such research, through new treatments or medical care. But Jacobs's XYY study, and ones that followed it, routinely neglected to include as subjects XYY males who were free and functioning normally in society. By focusing exclusively on inmate populations, these studies at least gave the appearance of searching for genetic confirmation that there was a biological basis for their subjects' powerlessness. At the same time XYY researchers tended to ignore or downplay a host of less exotic environmental factors — ranging from family history to childhood nutritional deficiency — that could conceivably have contributed to their subjects' mental disturbances.

THE SAME FEATURES that would later cause some scientists to openly question the value of XYY medical research made it an attractive science news story for the media. The possible link between heredity and crime has long been a topic of both scientific and public fascination. Here was a new opportunity to recycle that issue, in the light of the latest technological advances in genetics. Jacobs's 1965 survey was followed by a surge of similar studies that eagerly pursued the aggression hypothesis she had put forward in mental hospitals and penal institutions around the world. A number of them confirmed the disproportionately high rates of XYY males Jacobs had reported. But none succeeded in casting new light on any causal connection between chromosomes and criminality.

Nonetheless, these studies often received wide coverage in newspapers and magazines and on television. With disturbing frequency, the still highly preliminary XYY findings were publicly distorted and sensationalized. In the process, the media uncritically fostered an erroneous view of the supernumerary Y chromosome as the bearer of some sort of vaguely sinister "criminal genes."

During this period, *Newsweek*, for example, published a story on XYY males titled "Congenital Criminals." More dramatically, in 1968 the XYY story made national headlines in the United States with a rash of reports claiming that Richard Speck, the notorious criminal who had been convicted of murdering eight student nurses in Chicago, had an XYY genotype.

The reports were untrue and were later retracted. But they fed a growing public perception that populations of XYY men not only were genetically "driven" to live lives of violence and crime but also might represent a hidden menace to society. Some scientists even began to propose genetic screening programs designed to diagnose the XYY abnormality prenatally. Since there existed no medical remedy for XYY fetuses, the implication was that informed parents might simply choose to abort their unborn XYY males before these offspring could mature to act out their presumably genetically programmed "behavioral fate."

At first glance, the notion of systematically aborting babies bearing the XYY genotype must have struck some people as a eugenically sound plan — a pragmatic approach to gradually ridding the human gene pool of the newly defined XYY defect. If so, they were mistaken. Because the genetic missteps responsible for this trichromosomal state can occur anew in the meiotic cell divisions of each generation of parents, any such dreams of hereditary hygiene in the case of the XYY disorder would be illusory.

By 1968 defense attorneys in Australia and France had already tried to take advantage of this misconception by requesting reduced sentences for clients on trial for murder simply by showing them to be XYY. During the late 1960s, a U.S. government agency called the Center for Studies of Crime and Delinquency, a branch of the National Institute of Mental Health, concluded that the "XYY syndrome" fell under its broad mandate in the areas of crime, delinquency, and mental health. As a result it began to fund projects that, for example, carried out mass XYY screenings of juvenile males held in detention and treatment facilities in Maryland. And in 1969 it sponsored a major conference on XYY research, lending scientific legitimacy to the connection between XYY men and and crime.

By the early 1970s, a more disturbing trend had developed. Maternity hospitals in the United States, Canada, Denmark, and

England had begun to permit mass screening of newborn infants to see if they were XYY. On the surface such chromosomal testing, even for the enigmatic XYY defect, seems perfectly defensible. After all, the same karyotype that might identify an XYY baby could also reveal valuable information about more damaging genetic diseases. And even if the XYY genotype could not yet be conclusively linked to criminal aggression, it seemed that parents of an XYY baby at least had the right to know that such a possibility, however remote, existed.

ONE STUDY CARRIED OUT by Stanley Walzer, a Harvard child psychiatrist, and Park Gerald, a pediatrician, eventually became the focus of a heated public debate about the XYY controversy. In 1968, they had initiated an experimental genetic screening program at the Boston Hospital for Women that would produce within a day or two of birth karyotypes of male infants born in the maternity division. The purpose of the program was not only to identify boys with XYY genotypes but also to track them systematically for a period of years. Through periodic home visits, psychological tests and teacher questionnaires, the study would record each child's behavioral development in an effort to identify abnormalities.

Because no medical treatment existed for the XYY condition — no prescribed regimen of hormones, diet, or drugs that might alleviate any anticipated behavioral disorders — the researchers could not hope to treat XYY children displaying any of the expected signs of aggressive behavior. But they did propose to offer families what they termed "anticipatory guidance" — professional counseling sessions to help them cope with periodic behavioral problems their XYY boys might develop. In the end the hope was to amass enough information to be able to predict any XYY baby's risk of developing aggressive or sexual behavioral problems.

But the project would never achieve these ambitious aims. By 1974 a number of scientists and concerned private citizens had begun to challenge both the scientific validity and the ethical implications of the Boston XYY project, and they publicly called for it to be halted. The debate attracted considerable media attention and, in part because many scientists perceived it as a direct threat to their freedom to choose their own area of scientific research, quickly grew more rancorous. Among those leading the challenge were the geneticists Jonathan Beckwith of Harvard and Jonathan King of MIT, along with a Boston political organization called Science for the People.

In 1974, in an article in the British journal *New Scientist* titled "The XYY Syndrome: A Dangerous Myth," Beckwith and King outlined their objections to XYY research. From the beginning, they contended, the majority of XYY studies had been fundamentally flawed. They had been poorly designed, filled with logical inconsistencies and crippled by inadequate comparisons with matched, normally functioning XYY males as controls. The studies had often been precariously built on simplistic definitions of extremely complex human behaviors — including aggression, social deviance, and crime. And they had persistently downplayed the possible role of environmental factors in influencing traits too readily attributed to chromosomes.

But at the core of their critique was a concern for the serious ethical questions raised by a study such as Walzer's. First, researchers failed to obtain fully informed consent from participating parents. Because they were not told of the precise purpose of the karyotype tests or of the ambiguous medical findings from previous XYY studies, parents might not understand the dilemmas that could arise from an XYY diagnosis in their children. Second, if a child was diagnosed as XYY at birth and his family was informed of this diagnosis, that innocent child could be unfairly labeled at an early age as genetically prone to aggression and violence. This label could also contribute to a childhood setting in which a level of anger quite acceptable in a normal XY boy would be treated with undue concern by fearful parents aware of their child's genotype. In the end this distortion could generate new behavioral problems. In other words an XYY child could display abnormally aggressive behavior as a result of his role in the experiment — as the victim of a tragic self-fulfilling prophecy.

Beckwith filed a formal complaint with a Harvard research review committee, requesting that the body order a halt to the project on the grounds that it posed potential harm to its subjects and violated fundamental ethical principles of medical research. The committee eventually refused — but not before press conferences and public meetings brought the XYY issue into the public forum, earning additional support for the XYY critics. By 1975, faced with a growing public outcry over the ethical issues raised by his study, Walzer announced he would stop screening Boston newborns for the XYY genotype. Before long, scientists conducting a number of similar XYY studies in Denver and elsewhere decided to do the same — in part, to protect their patients from the public debate about their abnormality. Research on XYY males had simply become too hot to handle. And for the most part it has remained so to this day.

The climactic end of the Boston study closed a chapter in XYY research. But there remained many unanswered questions. Foremost was the mystery of the perplexingly high frequency of XYY males in high-security mental hospital settings. This mystery remains unsolved. At the same time, most XYY males — an estimated 96 percent — are thought to lead relatively ordinary lives, never seeing the inside of a prison or mental institution. Furthermore, medical experts, including Walzer and Gerald of the Boston XYY study, have now concluded that the list of supposed symptoms long thought to be associated with the XYY abnormality is so vague and inconsistent that the XYY genotype simply no longer merits the term "syndrome."

Tallness remains the most common physical characteristic found among XYY men. Some XYY males have facial skin problems, subtle speech and language difficulties, or childhood reading disabilities (all traits, incidentally, that could conceivably lead to aggressive or antisocial behavior in a young boy). Other XYY males are reported to have suffered from slight hormone imbalances or repeated outbursts of anger. But to date none of these signs is known to recur with regularity. And in many cases they could have easily arisen from causes other than an extra Y chromosome. In the words of a 1979 medical review of the XYY condition: "There is no characteristic that is uniformly present in patients with 47,XYY chromosome constitution other than the fact that they do have an extra Y chromosome."

IF THERE IS A SINGLE overriding lesson to be learned from the XYY debate, it goes far beyond the questions of the true clinical status, still largely unresolved, of XYY males, or even the unforeseen impact of XYY research on the lives of vulnerable human beings. Instead, this lesson must concern the incredible eagerness with which scientists and nonscientists alike have historically grasped at quick, socially convenient definitions of genetic disability or disease.

Overwhelmed by the complexities that underlie human differences, we are often too quick to judge one another on the basis of fragmentary genetic clues. In our impatience for easy answers to difficult questions, we run the risk of investing readily detectable hereditary traits with almost mystical prophetic powers. Like some prescientific shamanistic healer who finds people's fortunes revealed in fallen strands of hair or the discarded clippings of their fingernails, we unconsciously begin to extrapolate from the qualities of the part the qualities of the entire organism.

Regardless of which aspect of human genetic molecules we select for an early-warning system for human behavior — abnormalities in the number or structure of sex chromosomes, gene mutations or differences in DNA sequences so subtle that they leave no visible phenotypic spoor — we risk being deceived. Most geneticists agree that the vast majority of human behavioral traits tend to be polygenic; they reflect the simultaneous interaction of a multitude of genes. And every element in our species' rich behavioral repertoire is unavoidably sculpted at every stage of its development and action by a myriad of environmental variables too numerous to possibly anticipate.

To further complicate matters, the manifestation of even a single human gene can at times prove virtually impossible to predict. A given gene may simultaneously affect a number of seemingly unrelated traits. Or it may be expressed, depending on its surroundings, in a spectrum of possible phenotypes — some of them indistinguishable from those coded by entirely separate genes. In the face of these and other genuine biological complexities in human inheritance, how can we expect to reliably read a person's behavioral destiny in the tea leaves of an isolated hereditary flaw?

The XYY story may serve as a valuable example of what happens when people insist on simple answers to problems in human behavioral genetics and apply these provisional answers in too great haste.

But because the XYY story is unfinished, it can also leave us with a rather difficult personal dilemma. How are we supposed to react to the XYY genotype if it should, by chance, surface in our own individual lives? Should we be willing to welcome an unborn XYY child — however vague the suspected symptoms of his genotype — into the community of human beings?

Imagine, for instance, that you discover through prenatal tests that your unborn child is a boy and that, furthermore, he has an XYY genotype. Assuming your physician is knowledgeable about the XYY condition, he or she can only offer you uncertain counsel. Your son is likely to be born healthy and to lead a normal life, but some reports suggest that his extra Y chromosome may put him at slight statistical risk for developing subtle childhood behavioral or learning problems. The precise risk is unknown. The genetic mechanism is unknown. There is no reliable treatment. The choice of whether to proceed with the pregnancy is yours — the mother's and the father's. What choice would you make?

There are seldom easy answers or simple prescriptions for choices involving genetic defects and the life of an unborn child. Modern medicine has granted us increasing control over the size and the scheduling of our families. At the same time, it has placed the health of each planned child under closer medical scrutiny. Inevitably parents are driven to try to use this new knowledge to ensure that they will have a healthy, "normal" child. But prenatal screening tests seldom yield results that can be used to predict the health of an unborn child with absolute certainty.

In the case of our hypothetical XYY fetus, at least, that ambiguity is so extreme that the presence of an extra Y chromosome can scarcely even be referred to as a genetic disease. The condition is not known to seriously harm human health and has a dubious and an unpredictable effect on the bearer of the surplus chromosome. Yet it is not unlikely that many parents of an XYY fetus who have previously been exposed to media accounts of the highly publicized XYY controversy might opt for abortion — if only because of the vague, poorly documented but nonetheless disturbing association of the XYY genotype with aggression and crime.

Considering the medical literature available to date, we would find it difficult to condone such a response. In this case we find ourselves casting our hypothetical vote in favor of the chromosomally abnormal XYY child. (This is a strictly personal viewpoint; the decision, we believe, should be a matter of individual choice.) The reason: every one of us is genetically "flawed" to some extent. Our genomes are likely to be laced with all sorts of errors — mutations, recessive disease-causing alleles, miscopied genes, chromosomal blunders — most of them, for one reason or another, masked yet fully capable of surfacing in a subsequent generation. Some scientists have suggested that these genetic "defects" can actually be favored by natural selection. Chromosomal duplications, for instance, are thought to have played a key role in the molecular evolution of genes coding for the hemoglobin molecule found in human red blood cells.

Since none of us can hope to claim genetic "perfection" — in the face of an unpredictably changing environment, it is, after all, an absurd notion — how can we pretend to expect it in a newborn child? At least until such time as a particular genetic abnormality under consideration has been indisputably shown to place severe limits on a child's future health, happiness, or capacity to learn, it seems most wise to tolerate such imperfection, along with all the rest. The XYY genotype has not been shown to place such limits on a child's future. Therefore, it cannot be said to have passed this simple ethical test.

One suspects that a certain elderly and anonymous man — the man with the dubious distinction of being the first XYY man on record — would wholeheartedly agree. So might literally hundreds of thousands of XYY males living relatively normal lives at every level of society around the world (though few of them today could possibly know of their chromosomal blemish). So too, one might argue with tongue only partially in cheek, ought every male on Earth. For a compelling case could be made for declaring every genetically normal unborn male — equipped with a 46,XY genotype — at significant statistical risk for many of the same sinister characteristics that over the past quarter century have accumulated to stigmatize all 47, XYY males.

The reason: compared with their Y-less female counterparts, men — throughout history and in virtually all nations — have shown themselves to be statistically far more prone to lives of aggression, violence, and crime.

1990